T0140395

Cuatro Ciénegas Basin: An Endangered Hyperdiverse Oasis

Series Editors:

Valeria Souza
Ecology Institute
Universidad Nacional Autónoma de México
Mexico City, Distrito Federal, Mexico

Luis E. Eguiarte
Ecology Institute
Universidad Nacional Autónoma de México
Mexico City, Distrito Federal, Mexico

This book series describes the diversity, ecology, evolution, anthropology, archeology and geology of an unusually diverse site in the desert that is paradoxically one of the most phosphorus-poor sites that we know of. The aim of each book is to promote critical thinking and not only explore the natural history, ecology, evolution and conservation of the oasis, but also consider various scenarios to unravel the mystery of why this site is the only one of its kind on the planet, how it evolved, and how it has survived for so long.

More information about this series at http://www.springer.com/series/15841

Felipe García-Oliva • James Elser
Valeria Souza

Editors

Ecosystem Ecology and Geochemistry of Cuatro Cienegas

How to Survive in an Extremely Oligotrophic Site

 Springer

Editors
Felipe García-Oliva
Instituto de Investigaciones en Ecosistemas
y Sustentabilidad
Universidad Nacional Autónoma de México
Morelia, Michoacán, Mexico

James Elser
Flathead Lake Biological Station
University of Montana
Polson, MT, USA

Valeria Souza
Institute of Ecology
Universidad Nacional Autónoma de México
Mexico City, Distrito Federal, Mexico

ISSN 2523-7284   ISSN 2523-7292   (electronic)
Cuatro Ciénegas Basin: An Endangered Hyperdiverse Oasis
ISBN 978-3-030-07106-6   ISBN 978-3-319-95855-2   (eBook)
https://doi.org/10.1007/978-3-319-95855-2

This Springer imprint is published by the registered company Springer Nature Switzerland AG
The registered company address is: Gewerbestrasse 11, 6330 Cham, Switzerland

Preface

The Cuatro Ciénegas Basin (CCB) is a land of superlatives. For many, the super-latives focus on the valley's amazing biodiversity – the high level of endemism among its vertebrates, vascular plants, and, as revealed in Book 1, its astonishing microbiota. But the superlatives extend also to its physicochemical conditions and especially its biogeochemistry. This book seeks to provide a high-level over-view of work that has examined the chemical conditions found in both aquatic and terrestrial domains in the valley of Cuatro Ciénegas and of the ways that these conditions shape and are, in turn, shaped by, its biota.

One superlative that plays out in Cuatro Ciénegas ecosystems is extreme nutri-ent limitation. One sign of that nutrient limitation is in the crystal-clear waters of the valley's pozas and streams. Key nutrients such as nitrogen (N) and especially phosphorus (P) are at very low concentrations in CCB's waters, and thus they sup-port little biomass in the water column. However, much biomass is supported in the sediment and soil habitats and also in the diverse "living stromatolites" for which the valley is so famous. Chapters 1 and 2 describe, from a different perspec-tive, the nutrient conditions in Cuatro Ciénegas' soils and vascular plant commu-nities, highlighting, in what will become a recurring theme, the strong imbalance in the major chemical elements of carbon (C), nitrogen (N), and phosphorus (P) that are found there, and while N can be fixed and cycled by microbes (Chap. 2), C:N:P imbalance poses a significant challenge for soil fungi (Chap. 4). In Chap. 3, experiments manipulating N and P supply in the Churince drainage are described in which connections among microbial C:P and N:P ratios, RNA allocation, growth rate, and genomic traits are sought. In Chap. 5, a series of experiments demonstrating strong P limitation in stromatolite-based food webs is described and a picture emerges that, taking place in Rio Mesquites, is a form of P competi-tion between chemistry (calcification leading to coprecipitation of phosphate) and biology (growth of stromatolite microbes); chemistry seems to win, imposing severe P limitation on the food web. Building on the theme of P limitation, Chap. 9 explores how the life history of one microbial taxon (*Bacillus*) is associated, or not, with available P supplies. In another level of complexity, Chap. 11 reminds us that the aquatic consumers in Cuatro Ciénegas (snails, fishes) also have a role to play in the cycling of key nutrients in these oligotrophic systems. However, macronutrients such as N and P are not the whole story at Cuatro Ciénegas (or anywhere, for that matter). Thus, Chaps. 5 and 6 explore the sulfur cycle and its role in microbial biodiversity. Finally, we see fascinating data related to the ancient genes connected to iron playing a central role in Chap. 8 (magnetotactic bacteria) and in Chap. 10 (siderophores).

A major theme running through the chapters of this book and indeed through the chapters of Book 1 is that the unique ecosystems of Cuatro Ciénegas harbor unique microbiota that are intimately connected to the conditions formed by their desert environment. At Cuatro Ciénegas, that environment itself is tightly connected to the surrounding human community. In a desert region, water is quite obviously at a premium. Thus, changes in water use and distribution can have profound impacts. At Cuatro Ciénegas, these impacts are often immediate and obvious. Chapter 7 describes how the hyperdiverse microbial communities in the Churince system were disrupted by water depletion following agricultural diversion and extractions. Finally, in Chap. 12, we see signs of another human legacy: the impacts of agricultural practices on the normally nutrient-deficient soils.

Cuatro Ciénegas is a waterscape experiencing profound and diverse pressures, pressures that threaten the continued existence of this suite of ecologically unique ecosystems that have been evolving for many millennia. The physicochemical conditions found in CCB are of broad interest: the strong stoichiometric imbalance represents one of the most severely P-limited environments known, the stromatolite-based food webs serve as analogues of Earth systems during the Precambrian/Cambrian transition, and the valley's SO_4-dominated waters may represent conditions similar to those found on Mars as that planet lost its liquid water. Likewise, the microbes that have adapted to these unique conditions represent a storehouse of genomic innovation that we have only begun to explore. These unique treasures compel protection of these waters for future generations.

May the knowledge recorded in these chapters contribute to a strong foundation for that conservation.

Agua es vida.

Polson, MT, USA   James Elser
Morelia, Mexico   Felipe Garcia-Oliva
Coyoacan, Mexico   Valeria Souza

Contents

Chapter 1
Carbon, Nitrogen, and Phosphorus in Terrestrial Pools: Where Are the Main Nutrients Located in the Grasslands of the Cuatro Ciénegas Basin?

Felipe García-Oliva, Yunuen Tapia-Torres, Cristina Montiel-Gonzalez, and Yareni Perroni-Ventura

Contents

Abstract Photosynthesis is the process by which plants absorb atmospheric carbon (C) as they grow and convert it to biomass. However, plants acquire nitrogen (N) and phosphorus (P) only when these are available in the soil solution, which makes these elements the most limiting nutrients to plant growth and productivity in most terrestrial ecosystems. This chapter discusses the C, N, and P reservoirs in soil and plant biomass of two sites (Churince and Pozas Azules) in desert grassland

F. García-Oliva (✉) · C. Montiel-Gonzalez
Instituto de Investigaciones en Ecosistemas y Sustentabilidad, Universidad Nacional Autónoma de México, Morelia, Mexico

Y. Tapia-Torres
Escuela Nacional de Estudios Superiores Unidad Morelia, Universidad Nacional Autónoma de México, Morelia, Mexico
e-mail: ytapia@enesmorelia.unam.mx

Y. Perroni-Ventura
Instituto de Biotecnología y Ecología Aplicada, Universidad Veracruzana, Xalapa, Mexico

© Springer International Publishing AG, part of Springer Nature 2018
F. García-Oliva et al. (eds.), *Ecosystem Ecology and Geochemistry of Cuatro Cienegas*, Cuatro Ciénegas Basin: An Endangered Hyperdiverse Oasis, https://doi.org/10.1007/978-3-319-95855-2_1

1

dominated by *Sporobolus airoides* at the Cuatro Ciénegas Valley. We also analyzed the influence of different factors such as soil nutrients, water availability, and microbial nutrient transformations that determine the resource allocation to different pools in this oligotrophic ecosystem. We observed higher aboveground and belowground biomass in Churince than in Pozas Azules. Additionally, we observed higher C and P contents in roots, higher soil total organic C and organic P at Churince, and higher N concentration in the aboveground grass biomass at Pozas Azules. Nutrient contents showed different patterns between sites. Total carbon, N, and P contents were all higher in Churince than Pozas Azules. At the ecosystem level, organic C and organic P were higher in Churince, but no differences were observed in N. In the two soil types studied, C:N:P stoichiometric ratios were different, suggesting that the same dominant plant species makes different adjustments of nutrient concentrations depending on water and nutrient availability, a response that can affect ecosystem nutrient pools as well as various ecosystem processes.

Keywords C:N:P stoichiometric ratios · Ecosystems pools · grassland · Nutrients dynamic · Soil

Introduction

Grasslands represent 32% of the continental surface of Earth (Adams et al. 1990). Indeed, arid and semiarid ecosystems store twice as much carbon (C) as temperate forest ecosystems (Taylor and Lloyd 1992). Lal (2009) estimated that 15% of the global soil organic carbon is stored in dryland (arid and semiarid) ecosystems. Therefore, C fluxes in arid ecosystems may strongly affect the global C cycle, especially during rainy years (Poulter et al. 2014). For example, Australian grasslands contributed 51% of the global C sink in the year 2011 as a consequence of La Niña rainy conditions during that year (Poulter et al. 2014). These results suggest that arid grasslands can play an important role as a C sink during wet years. Supporting this, Petrie et al. (2015) executed a watering experiment in the Chihuahuan Desert and found that the grassland had net C uptake in wet conditions but was a net C emitter in dry conditions.

However, grassland productivity is not only affected by water availability but also can be constrained by soil nutrient availability, particularly nitrogen (N; Burke et al. 1998; Whitford 2002) and phosphorus (P; Elser et al. 2007; Jouany et al. 2011; Tapia-Torres et al. 2015a, b). Indeed, the C, N, and P cycles are biologically coupled, and this coupling establishes the elementary stoichiometry of plant and soil microbial communities (Finzi et al. 2011). Elser et al. (2000) reported that terrestrial autotrophs, mainly plants, have an average C:N:P ratio of 968:28:1, while the average C:N:P ratio of soil microbial biomass is 60:7:1 (Cleveland and Liptzin 2007). This stoichiometric contrast suggests that the plants are nutrient-poor in comparison with the soil microbial community, and, therefore, investments in acquiring nutrients must be different for the two components. As a consequence, assessing the distribution of the elements (C, N, and P) among ecosystem pools may aid in understanding how these nutrients are available to different organisms within the ecosystem.

The distribution of nutrients among ecosystems pools is critical for understanding their vulnerability to natural or anthropogenic disturbances. In general, soil nutrient pools are more protected over mid- and long-term time scales than those within biomass, especially that aboveground (García-Oliva et al. 2006). Around 60% of ecosystem C pools in tropical rain forests are concentrated within aboveground biomass (García-Oliva et al. 2006; Hughes et al. 2000; Trumbore et al. 1995), while in temperate forest ecosystems, more than 60% of C is stored in the soil (Gallardo and González 2004; García-Oliva et al. 2006; Ordoñez et al. 2008). Consequently, ecosystem C is very vulnerable to deforestation in tropical rain forests. For example, Hughes et al. (2000) reported that 50% of C ecosystem content of tropical rain forest is lost by deforestation per year. While there are many studies of ecosystem C pools in different ecosystems worldwide, there are far fewer studies report N and P ecosystem pools.

As mentioned before, the productivity of arid grassland strongly fluctuates with annual rainfall (Chen et al. 2009; Petrie et al. 2015; Poulter et al. 2014). However, the main organic C input to the soil in this ecosystem occurs belowground rather than aboveground (Austin et al. 2004). In arid and semiarid grassland in Mexico, the belowground biomass has three times higher C content than the aboveground biomass (Montaño et al. 2016). These results suggest that the more stable C pools in these ecosystems are in the soil. Therefore, the main objective of this chapter is to summarize our efforts to quantify of C, N, and P contents within the three main ecosystem pools (above- and belowground plant biomass and soil) in two grassland sites with different soil types and water availability in the Cuatro Ciénegas Valley. Also, we analyzed C:N:P stoichiometric ratios of the different ecosystem pools.

Materials and Methods

Study Area

This research was carried out in the Cuatro Ciénegas Valley in the north of Mexico, in the central Chihuahuan Desert (26° 50′ 41″ N and 102° 8′ 11″ W). This valley occupies an area of 30 × 40 km, and the bottom of the valley lies at 740 m above sea level. The climate is seasonally arid with an average annual temperature of 21 °C and a mean annual rainfall of 252 mm, which is concentrated during the summer months (http://smn.cna.gob.mx/) (See Chapter 2, climate chapter Book 1). The valley is surrounded by limestone ridges and separated into two lobes by the San Marcos Ridge. The eastern side of the valley is dominated by Cretaceous carbonate limestone, and the western side is dominated by Jurassic gypsum (McKee et al. 1990). As a consequence, the dominant soils in the basin are *calcisols* and *gypsisols* in the east and west side, respectively (Tapia-Torres et al. 2015b), according to IUSS Working Group WRB (WRB 2007). The dominant vegetation in the bottom of the valley is grassland dominated by *Sporobolus airoides* (Torr.) Torr. (Pinkava 1974; Perroni et al. 2014a).

Sampling

To compare the ecosystem nutrient pools of grassland in these two different soil types, we selected one site for each soil type: Churince, which is dominated by gypsisols (west side; Ch), and Pozas Azules, which is dominated by calcisols (east side; PA). At each site, a random 50 × 100 m plot was established within grassland. At each plot, ten transects were measured every 10 m, and a sampling plot was randomly chosen (2 × 2 m size). Each sampling plot was divided in four subplots for sampling aboveground biomass of grass. In two of the subplots of each sampling plot, all aboveground biomass was collected and stored in black plastic bags until analysis in the laboratory. Similarly, five soil samples (four corners and one central point) were composited from each sampling plot. Additionally, one soil sample was obtained for each sampling plot for the determination of soil bulk density. Soil samples were taken from the top 15 cm depth of mineral soil. Soil was placed in black plastic bags and kept at 4 °C until laboratory analyses were carried out. Root samples were collected in each of the 10 transects of 50 m length, covering the entire sampling area (five samples per line; $n = 50$). Root samples were collected similarly to soil samples and stored in the same way until laboratory analysis.

Laboratory Analyses

The aboveground biomass and soil samples used to determine root biomass were dried at 60 °C until they reached a constant weight. The aboveground samples were weighed and ground for chemical analyses. The soils of root samples were sieved through different sieve sizes (No. 8 of 2.3 mm, No. 20 of 0.84 mm, and No. 45 of 335 µm), and roots were removed with tweezers and a magnifying glass. Roots obtained were dried at 30 °C, weighed, and then ground for chemical analyses. Vegetative material was homogenized in a mill (Thomas Scientific) to pass through a 40 mesh screen (0.425 mm). Total C was determined by coulometric detection (Huffman 1977). Total N and P were determined after acid digestion. Nitrogen was determined by a macro-Kjeldahl method adapted for vegetative material (Bremmer 1996), and P was determined by the molybdate colorimetric method after ascorbic acid reduction (Murphy and Riley 1962).

The nutrient concentration data from the Ch soil are reported in Tapia-Torres et al. (2015a). The PA soil samples were analyzed with the same methods used for the Ch soil samples. Details of methods used are described in Tapia-Torres et al. (2015a).

Briefly, a soil subsample was oven-dried at 105 °C to constant weight for soil moisture determination. Each dry sample was ground in an agate mortar prior to total soil nutrient analyses. Total C and inorganic C were determined by dry combustion and coulometric detection using the Carbon Analyzer (UIC Mod. CM5012; Chicago, USA; Huffman 1977). Organic C was calculated as the difference between total and inorganic C. Total N (Nt) was determined after acid digestion by the macro-Kjeldahl method and determined colorimetrically (Bremmer 1996). Total P was determined by the molybdate colorimetric method after ascorbic acid reduction (Murphy and Riley 1962). Total soil P in each site was not statistically different from the sum of the all soil fractions obtained by Hedley sequential extraction method, as reported by Perroni et al. (2014b) for the Ch soil and by Montiel (2011) for the PA soil. Thus, total inorganic P (Pi) and organic P (Po) were obtained by the multiplication of total P by the percentage of the sum of inorganic or organic fraction estimated by the sequential extraction method in Perroni et al. (2014b) and Montiel (2011).

Data Analyses

The total contents for C, N, and P in above- and belowground grass biomass were estimated by multiplying the respective biomass (above- or belowground) by their respective concentration (mass per dry mass) in each biomass pool. Soil nutrient contents were calculated by multiplying the soil mass (estimated with bulk density and soil depth sampling) by the concentrations of each nutrient. The total ecosystem pools were calculated by adding aboveground, belowground, and soil nutrient contents.

Differences in nutrient concentrations, nutrient ratios, and nutrient contents between grassland sites were analyzed with a Student's t-test. A linear regression was performed to test if soil available or microbial nutrient concentrations correlated with nutrient concentrations in *S. airoides* biomass. All statistical analyses were performed using Statistica 7.0 (StatSoft).

Results

Nutrients in Above- and Belowground Grass Biomass

Aboveground grass biomass in Churince was higher than that in Pozas Azules (560 ± 69 and 323 ± 93 gm^{-2}, respectively); this was also true for belowground biomass (684 ± 155 and 288 ± 31 gm^{-2} for Ch and PA, respectively). As a result, total grass biomass in Ch was two times higher than in PA (1244 ± 154 and 611 ± 176 gm^{-2}, respectively). Table 1.1 shows grass biomass nutrient concentrations within above- and belowground biomass. PA grassland had higher C and N concentrations in aboveground biomass than the Ch grassland. Additionally, PA had

Table 1.1 Average (with standard errors) of grass and soil nutrient concentrations in two grassland sites within Cuatro Ciénegas Basin with Student's t-test results

Variable	Churince	Pozas Azules	t (p)
Grass nutrient concentrations (mg g^{-1})			
Aboveground C	370 (3)	382 (5)	−2.6 (0.05)
Aboveground N	4.3 (0.26)	6.5 (0.28)	−5.8 (<0.001)
Aboveground P	0.33 (0.05)	0.24 (0.01)	1.9 (0.07)
Belowground C	293 (7)[a]	197 (9)	8.5 (<0.001)
Belowground N	10.7 (0.8)[a]	9.4 (1.3)	0.9 (0.38)
Belowground P	0.50 (0.05)[a]	0.24 (0.01)	5.5 (<0.001)
Soil nutrient concentrations			
Total soil nutrients (mg g^{-1})			
Organic C	13.4 (1.8)[a]	5.9 (0.7)	3.8 (0.001)
Nitrogen	0.89 (0.17)[a]	0.63 (0.07)	1.4 (0.16)
Inorganic P	0.044 (0.01)[b]	0.076 (0.01)	−3.3 (<0.001)
Organic P	0.043 (0.01)[b]	0.018 (0.00)	4.0 (<0.001)

[a]Data from Tapia-Torres et al. (2015a)
[b]Data from Perroni et al. (2014b)

Table 1.2 Average (with standard errors) of grass, soil, and ecosystem nutrient ratios in two grasslands within the Cuatro Ciénegas Basin with Student's t-test results

Variable	Churince	Pozas Azules	t (p)
Grass nutrient ratios			
Aboveground C:N	88 (5)	59 (3)	4.9 (<0.001)
Aboveground C:P	1349 (188)	1668 (93)	−1.5 (0.14)
Aboveground N:P	15 (2)	26 (2)	−4.8 (<0.001)
Belowground C:N	29 (2)	24 (3)	1.4 (0.17)
Belowground C:P	628 (66)	821 (48)	−2.4 (0.02)
Belowground N:P	24 (4.5)	39 (6.0)	−2.0 (0.06)
Soil nutrient ratios			
Total organic C:N	17 (1.9)	9.3 (0.4)	3.9 (<0.001)
Total organic C: Pi	313 (31)	80 (6)	7.4 (<0.001)
Total organic C:Po	308 (30)	330 (25)	−0.5 (0.59)
Total N:Pi	20 (2)	9 (1)	5.5 (<0.001)
Total N:Po	20 (2)	36 (3)	−4.5 (<0.001)
Ecosystem nutrient ratios			
Organic C:N	18 (2)	10 (1)	4.1 (<0.001)
Organic C:Po	329 (27)	366 (28)	−0.9 (0.36)
N:Po	19 (2)	35 (3)	−4.6 (<0.001)

lower C and P concentrations in belowground biomass than the Ch grassland. Therefore, the C:N ratio of aboveground biomass was lower in PA than in Ch (Table 1.2), but the N:P ratio of the aboveground biomass and the C:P ratio of belowground biomass were lower in the Ch grassland than in PA (Table 1.2). However, the aboveground biomass contents of C, N, and P were not significantly

Table 1.3 Average (with standard errors) of plant, soil, and ecosystem nutrient content in two grasslands within the Cuatro Ciénegas Basin with Student's *t*-test results

Variable	Churince	Pozas Azules	t (p)
Grass nutrient contents (gm^{-2})			
Aboveground C	182 (22)	124 (36)	1.3 (0.19)
Aboveground N	2.15 (0.3)	1.97 (0.5)	0.3 (0.74)
Aboveground P	0.17 (0.04)	0.08 (0.03)	1.9 (0.06)
Belowground C	222 (50)	58 (6)	3.2 (0.004)
Belowground N	7.6 (1.6)	2.7 (0.4)	2.9 (0.008)
Belowground P	0.55 (0.11)	0.07 (0.01)	4.4 (<0.001)
Soil nutrient contents (gm^{-2})[a]			
Organic C	2804 (364)	1173 (141)	4.2 (<0.001)
Nitrogen	186 (33)	125 (13)	1.7 (0.09)
Inorganic P	9.2 (1.1)	14.9 (1.6)	−3.0 (0.007)
Organic P	9.3 (1.1)	3.6 (0.4)	4.9 (<0.001)
Ecosystem nutrient contents (gm^{-2})			
Organic C	3209 (371)	1356 (161)	4.6 (<0.001)
Nitrogen	196 (33)	130 (13)	1.9 (0.07)
Organic P	10.0 (1.1)	3.8 (0.4)	5.1 (<0.001)

[a]Top 15 cm soil depth

different between the two grassland sites, while belowground biomass C, N, and P contents were higher in the Ch grassland than in the PA grassland (Table 1.3).

The aboveground/belowground C content ratio exceeded one for both sites, and there were no statistical differences between sites (Student's *t*-test: −0.6, $P = 0.56$; Fig. 1.1), indicating that both sites had more aboveground C than belowground C. In contrast, the belowground N content in both sites and the P content in Ch grassland were higher than the aboveground content. As a consequence, the aboveground/belowground content ratios were lower than one (Fig. 1.1). The aboveground/belowground N content ratio did not differ for the two sites (*t*: −1.2, $P = 0.24$; Fig. 1.1), but the Ch grassland had lower aboveground/belowground P content ratio than the PA grassland (*t*: −2.7, $P = 0.01$; Fig. 1.1).

Soil Nutrient and Ecosystem Nutrient Contents

Ch soil had higher moisture than the PA soil (37% ± 2 vs 22% ± 1, respectively; *t*: 6.3, $P < 0.001$), while soil pH was higher in PA than in Ch (9.0 ± 0.1 and 8.6 ± 0.1, respectively; *t*: −4.98, $P < 0.001$). Total organic C (OC) and total organic P (Po) concentrations were higher in Ch soil than in PA soil, but the latter had higher total inorganic P (Pi) than the former (Table 1.1). As a result, the OC:N, OC:Pi, and N:Pi ratios were higher in Ch soil than in PA soil, but PA had higher N:Po ratio than Ch (Table 1.2). Finally, Ch soil had higher OC and Po contents than PA soil but had lower soil Pi contents than PA (Table 1.3).

Fig. 1.1 Aboveground/
belowground ratios of
nutrient contents in two
grasslands within Cuatro
Ciénegas Basin. *: means
that the P ratios were
statistically different
between the two sites
(*P* = 0.01)

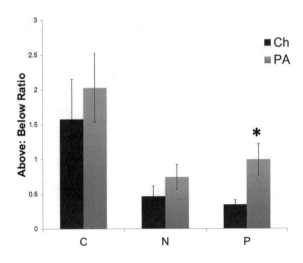

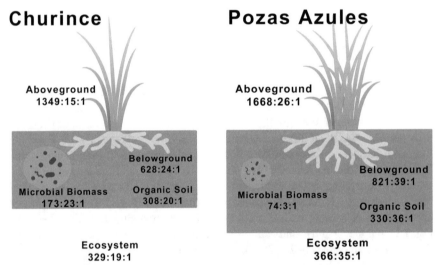

Fig. 1.2 Graphic representations of C:N:P ratios of the principal pools in two grassland sites at Cuatro Ciénegas Valley. Higher values indicate that this pool had lower P. The C:N:P ratios of microbial biomass are reported by Tapia-Torres et al. (2015b). The size of the pools are according to C:P ratios (aboveground biomass, soil, and microbial biomass)

Whole ecosystem C and Po contents were higher in Ch than in PA, while ecosystem N content did not differ for the two sites (Table 1.3). Therefore, the OC:N ratio was higher at Ch site, while the N:Po ratio was higher at PA (Table 1.2). Figure 1.2 shows the C:N:P ratios of the main ecosystem pools of both sites. Churcince had lower values for most of the ecosystem pools than Pozas Azules with the exception of microbial biomass, which was higher at PA. Figure 1.3 shows the relative distribution of nutrient contents within the ecosystems, where the soil (the top 15 cm depth) is the main pool of organic C, N, and P (Fig. 1.3).

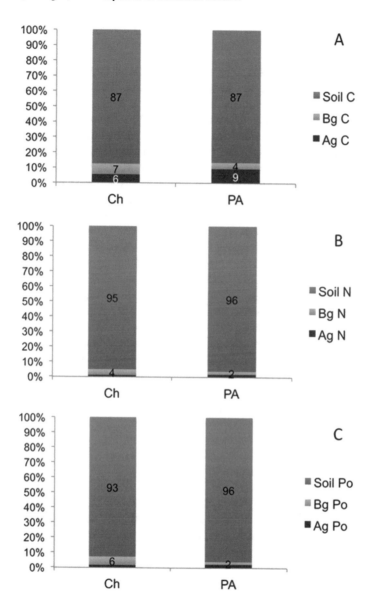

Fig. 1.3 Relative distribution of nutrient contents among aboveground biomass (Ag), below-ground biomass (Bg), and soil (top 15 cm depth soil) of (**a**) carbon, (**b**) nitrogen, and (**c**) phosphorus in two grasslands within the Cuatro Ciénegas Basin. The numbers within column represent the percentage of each ecosystem pool

Discussion

Soil water availability seems to be an important factor that promotes higher grass biomass productivity in the wet site (Churince), as in other arid and semiarid grass-lands (Chen et al. 2009; Hao et al. 2008; Petrie et al. 2015). However, our results of nutrient concentrations and nutrient ratios within grass biomass suggest that grasses at the two sites exhibit different nutrient use and nutrient uptake strategies. Aboveground grass biomass in the dry site (PA) had higher N concentration than the wet site (Ch; 6.5 vs 4.3 mg g^{-1}, respectively) and lower C:N ratio (59 vs 88, respec-tively). The C:N ratio of aboveground biomass of the PA grassland (59) is similar to the C:N ratio of tree foliage in temperate coniferous forests (59) reported by Cleveland and Liptzin (2007). However, the C:N ratios of aboveground biomass from both sites are higher than the reported average of plant biomass (36) by Elser et al. (2000). These results suggest that these grasses within CCB are N-limited in comparison with other plant communities, especially at the wet site (Ch; C:N = 88). According to Sterner and Elser (2002), high values of C:N and C:P ratios in auto-troph biomass suggest potential N- and/or P-limitation of growth. In contrast, the N:P ratio of aboveground biomass was higher in the dry site than in the wet site (26 and 15 for PA and Ch, respectively). These values indicate that the grasses at the wet Ch site had a more effective strategy of P acquisition. Additionally, these N:P ratios are similar to or lower than the average N:P ratio of plant biomass (28) reported by Elser et al. (2000) and also lower for the N:P ratio for average of tree foliage of tropical and temperate forests (28) reported by Cleveland and Liptzin (2007). N:P ratios of plant leaves below 10 are indicative of N-limitation, while N:P ratios higher than 20 indicate P-limitation (Güsewell 2004). Thus, our results suggest that the grasses in the wet site (Ch) are more limited by N, while the grasses of the dry site (PA) are more limited by P. Similarly, Tapia-Torres et al. (2015b) reported that the soil microbial community is more limited by P in the soil of the dry site (PA), while the microbial community in the soil of the wet site (Ch) is more N-limited.

C content was slightly higher in aboveground biomass than in belowground bio-mass. However, the grasses allocated more N in both sites and more P in the wet site (Ch) to belowground biomass. Also, P in the dry site (PA) is equally allocated to above- and belowground biomass. These results suggest that these plants allocated more nutrients to root biomass to acquire soil nutrients, as reported before (Aerts and Chapin III 1999). The higher allocation of N and P to root biomass suggested that the plants, especially in the wet site, are invested in proteins and ATP in roots to promote active nutrient uptake, production of exudates, and microbial symbiotic associations (George et al. 2011; Raghothama 1999; Zhang et al. 2014). For exam-ple, Hernandez-Hernández (pers. comm.) found that the grasses of the dry site (PA) exhibited higher arbuscular mycorrhizal fungi infection and external mycelium pro-duction than the grasses of the wet site (Ch) during a dry year where the soil P was less available for the microbial community (Tapia-Torres et al. 2015b) and probably for the plants. Additionally, the wet site (Ch) had lower C:N:P ratios in the majority of ecosystem pools than the dry site (PA), with the exception of the soil microbial biomass. The results confirm that the dry site (PA) is more limited by P than the wet site (Ch). However, Tapia-Torres et al. (2016) found that the soil bacterial community has the capacity of release P from different organic and inorganic molecules and use P forms in different oxidation states. For example, bacteria isolates can use P via mineralization of phosphonates (a stable organic molecule), P solubilization of calcium phosphate, and possible oxidation of phosphite (Tapia-Torres et al. 2016). Phosphite may be generated in some ecosystems by the degradation of different types of phosphonates (Pasek et al. 2014). The presence of bacterial species with different strategies of P acquisition increases the availability of P and therefore promotes its flow through the different ecosystems pools.

Our results suggest that organic soil compounds are the main pools of ecosystem C, N, and P in both sites within the CCB. This distribution makes the ecosystem less vulnerable to nutrient losses, as the nutrients are recycled within biological feedbacks. However, reductions of water availability, and therefore reductions of input of organic C to the soil, favors nitrification over denitrification, which promotes ecosystem soil N losses (López-Lozano et al. 2012; Tapia-Torres et al. 2015a). Therefore, we suggest that the resilience of nutrient contents in this semiarid ecosystem strongly depends on organic matter input to the soil, as well as the soil microbial community composition, and ultimately on water availability.

Acknowledgments We thank Rodrigo Velazquez-Duran for his assistance during chemical analyses. We also thank the personnel of APFF Cuatro Ciénegas (CONANP) and the people in charge of Rancho Pozas Azules (PRONATURA) for permission to collect soil samples on their respective properties. This work was financed by the National Autonomous University of Mexico (PAPIIT DGAPA-UNAM grant to FGO: El papel de la disponibilidad del Carbono sobre la dinámica del Nitrógeno y Fósforo edáfico en ecosistemas contrastantes de México, IN201718).

References

Adams JM, Faure H, Faure-Denard L et al (1990) Increases in terrestrial carbon storage from the Last Glacial Maximum to the present. Nature 348:711–714

Aerts R, Chapin FS III (1999) The mineral nutrition of wild plants revisited: a re-evaluation of processes and patterns. Adv Ecol Res 30:1–67

Austin AT, Yahdjian L, Stark JM et al (2004) Water biogeochemical pulses and cycles in arid and semiarid ecosystems. Oecologia 141:221–235

Bremmer JM (1996) Nitrogen-Total. In: Spark DL, Page AL, Summer ME, Tabatabai MA, Helmke PA (eds) Methods of soil analyses part 3: chemical analyses. Soil Science Society of America, Madison, WI, pp 1085–1121

Burke IC, Lauenroth WK, Vinton MA et al (1998) Plant-soil interactions in temperate grasslands. Biogeochemistry 42:121–143

Chen S, Lin G, Huang J, Jenerette GD (2009) Dependence of carbon sequestration on the differential responses of ecosystem photosynthesis and respiration to rain pulses in a semiarid steppe. Glob Chang Biol 15:2450–2461

Cleveland CC, Liptzin D (2007) C:N:P stoichiometry in soil: is there a "Redfield ratio" for the microbial biomass? Biogeochemistry 85:235–252

Elser JL, Fagan WF, Denno RF et al (2000) Nutritional constraints in terrestrial and freshwater foodwebs. Nature 408:578–580

Elser J, Bracken MES, Cleland EE et al (2007) Global analysis of nitrogen and phosphorus limitation of primary producers in freshwater, marine and terrestrial ecosystems. Ecol Lett 10:1135–1142

Finzi AC, Austin AT, Cleland EE et al (2011) Coupled biochemical cycles: responses and feedbacks of coupled biogeochemical cycles to climate change. Examples from terrestrial ecosystems. Front Ecol Environ 9:61–67

Gallardo J, Gónzalez MI (2004) Sequestration of carbon in Spanish deciduous oak forests. Adv Geogr Ecol 37:341–351

García-Oliva F, Hernández G, Gallardo JF (2006) Comparison of ecosystem C pools in three forests in Spain and Latin America. Ann For Sci 63:519–523

George TS, Fransson AM, Hammond JP, White PJ (2011) Phosphorus nutrition: Rhizosphere processes, plant response and adaptations. In: Bünemann EK, Oberson A, Frossart E (eds) Phosphorus in action: biological processes in soil phosphorus cycling. Springer-Verlag, Berlin, Heidelberg, pp 245–271

Güsewell S (2004) N:P ratios in terrestrial plants: variation and functional significance. New Phytol 164:243–266

Hao Y, Wang Y, Meid X et al (2008) CO_2, H_2O and energy exchange of an Inner Mongolia steppe ecosystem during a dry and wet year. Acta Oecol 33:133–143

Huffman EWD (1977) Performance of a new carbon dioxide coulometer. Microchem J 22:567–573

Hughes RF, Kauffman JB, Jaramillo VJ (2000) Ecosystem-scale impacts of deforestation and land use in a humid tropical region of Mexico. Ecol Appl 10:515–527

IUSS Working Group WRB (2007) World reference base for soil resources, first update 2007. World soil resources reports no. 103. FAO, Rome

Jouany C, Cruz P, Daufresne J, Duru M (2011) Biological phosphorus cycling in grasslands: interaction with nitrogen. In: Bünemann EK, Oberson A, Frossard E (eds) Phosphorus in action: biological processes in soil phosphorus cycling. Springer, Berlin, Heidelberg, pp 295–316

Lal R (2009) Sequestering carbon in soils of arid ecosystems. Land Degrad Dev 20:41–454

López-Lozano NE, Eguiarte LE, Bonilla-Rosso G et al (2012) Bacteria communities and nitrogen cycle in the gypsum soil in Cuatro Cienegas Basin, Coahuila: a Mars analogue. Astrobiology 12:699–709

McKee JW, Jones NW, Long LE (1990) Stratigraphy and provenance of strata along the San Marcos fault, central Coahuila, Mexico. Geol Soc Am Bull 102:593–614

Montaño NM, Ayala F, Bullock SH et al (2016) Almacenes y flujos de carbono en ecosistemas áridos y semiáridos de México: síntesis y perspectivas. Terra Latinoam 34:39–59

Montiel González C (2011) Dinámica de C, N y P en suelos calcáreos en el valle de Cuatro Ciénegas de Carranza, Coahuila. Master Dissertation, Universidad Nacional Autónoma de México

Murphy J, Riley JP (1962) A modified single solution method for the determination of phosphate in natural waters. Anal Chim Acta 27:31–36

Ordoñez JAB, de Jong BHJ, García-Oliva F et al (2008) Carbon content in vegetation, litter, and soil under 10 different land-use and land-cover classes in the Central Highlands of Michoacan, Mexico. For Ecol Manag 255:2074–2084

Pasek MA, Sampson JM, Atlas Z (2014) Redox chemistry in the phosphorus biogeochemical cycle. PNAS 111:15468–15473

Perroni Y, García-Oliva F, Souza V (2014a) Plant species identity and soil P forms in an oligotrophic grassland–desert scrub system. J Arid Environ 108:29–37

Perroni Y, García-Oliva F, Tapia-Torres Y et al (2014b) Relationship between soil P fractions and microbial biomass in an oligotrophic grassland-desert scrub system. Ecol Res 29:463–472

Petrie MD, Collins SL, Swann AM et al (2015) Grassland to shrubland state transitions enhance carbon sequestration in the northern Chihuahuan Desert. Glob Chang Biol 21:1226–1235

Pinkava DJ (1974) Vegetation and flora of the Bolson of Cuatro Ciénegas Region, Coahuila, Mexico: IV. Summary, endemism and corrected catalogue. J Ariz Nev Acad Sci 19:23e47

Poulter B, Frank D, Ciais P et al (2014) Contribution of semi-arid ecosystems to interannual variability of the global carbon cycle. Nature 509:600–604

Raghothama KG (1999) Phosphate acquisition. Ann Rev Plant Physiol Plant Mol Biol 50:665–693
Sterner RW, Elser JJ (2002) Ecological stoichiometry: the biology of elements from molecules to the biosphere. Princeton University Press, Princeton, NJ
Tapia-Torres Y, López-Lozano NE, Souza V et al (2015a) Vegetation-soil system controls soil mechanisms for nitrogen transformation in a oligotrophic Mexican desert. J Arid Environ 114:62–69
Tapia-Torres Y, Elser JJ, Souza V et al (2015b) Ecoenzymatic stoichiometry at the extremes: how microbes cope in a ultraoligotrophic desert soil. Soil Biol Biochem 87:34–42
Tapia-Torres Y, Rodríguez-Torres MD, Islas A, Elser J et al (2016) How to live with phosphorus scarcity in soil and sediment: lessons from bacteria. Appl Environ Microbiol 82:4652–4662
Taylor JA, Lloyd J (1992) Sources and sinks of atmospheric CO_2. Aust J Bot 40:407–418
Trumbore SE, Davison EA, Barbosa de Carmargo P et al (1995) Belowground cycling of carbon in forest and pasture of Eastern Amazonia. Glob Biogeochem Cycles 9:515–528
Whitford WG (2002) Ecology of desert systems. Academic Press, London, UK
Zhang Z, Liao H, Lucas W (2014) Molecular mechanisms underlying phosphate sensing, signaling, and adaptation in plants. J Integr Plant Biol 56:192–220

Chapter 2
Terrestrial N Cycling in an Endangered Oasis

Nguyen E. López-Lozano, Ana E. Escalante, Alberto Barrón-Sandoval, and Teresa Perez-Carbajal

Contents

Abstract In terrestrial arid ecosystems, one of the most limiting factors for productivity, following water, is thought to be nitrogen (N) availability. The N cycle can be summarized as an exchange of N forms between the atmosphere and the biosphere, mediated by the biological activity of microorganisms. Arid lands typically have a heterogeneous distribution of resources, with vegetated areas and microbial crusts having greater nutrient concentrations and microbial densities than bare soils. However, the contribution of each compartment to the entire N budget in these arid ecosystems is poorly understood. In this chapter, we summarize studies performed in the terrestrial component of Cuatro Cienegas Basin (CCB) regarding different aspects of the N cycle. We present selected results from two different studies that contrast microbial diversity and specific N transformations in (i) different moisture conditions and (ii) different temperatures. Although microbial crusts are important components of many desert ecosystems, there is very little evidence that the N fixed within them is in turn

N. E. López-Lozano (✉)
CONACyT-División de Ciencias Ambientales, Instituto Potosino de Investigación
Científica y Tecnológica (IPICYT), San Luis Potosi, Mexico
e-mail: nguyen.lopez@ipicyt.edu.mx

A. E. Escalante · A. Barrón-Sandoval · T. Perez-Carbajal
Laboratorio Nacional de Ciencias de la Sostenibilidad, Departamento de Ecología de la
Biodiversidad, Instituto de Ecología, Universidad Nacional Autónoma de México,
Coyoacan, Mexico

© Springer International Publishing AG, part of Springer Nature 2018
F. García-Oliva et al. (eds.), *Ecosystem Ecology and Geochemistry of Cuatro Cienegas*, Cuatro Ciénegas Basin: An Endangered Hyperdiverse Oasis,
https://doi.org/10.1007/978-3-319-95855-2_2

available to higher plants. Considering this, N fixers in the rhizosphere of plants could also be relevant N suppliers. In the last part of this chapter, we compare the potential composition of the microbial N fixers and denitrifier communities present in bare soils and in the rhizosphere of *Agave lechuguilla*, one of the most characteristic plant species in the Mexican arid regions. In general, these data suggest that environmental changes such as soil moisture reduction, changes in temperature, and vegetation removal could dramatically affect the terrestrial N cycle in CCB.

Keywords Denitrifier communities · N fixers · Nitrification · Soil bacteria · Soil microbial crust

N Cycle in Arid Lands

In terrestrial arid ecosystems, one of the most limiting factors for productivity, following water, is thought to be nitrogen (N) availability. The N cycle can be summarized as an exchange of N forms between the atmosphere and the biosphere, with a network of oxidation–reduction reactions mediated by the biological activity of plants, fungi, bacteria, and archaea. These organisms modulate the oxidation state of N between that of fully reduced amines and fully oxidized nitrate. The largest pool of N in the biosphere is dinitrogen gas (N_2), which comprises around 78% of the atmosphere, but this is not directly available to most organisms with the exception of a small number of nitrogen-fixing Archaea and Bacteria (Zehr and Kudela 2011). Biological fixation is catalyzed by the enzyme nitrogenase (encoded by *nif* genes), which fixes N_2 as biologically available ammonium (NH_4) that then breaks down to NH_3 plus H^+, allowing NH_3 to be assimilated by living things (Devol 1991). Ammonium can also be readily oxidized by soil microbes, producing hydroxylamine (NH_2OH), nitrite (NO_2), and nitrate (NO_3). This process is catalyzed by microorganisms termed ammonia-oxidizing bacteria and archaea (AOB and AOA, respectively), containing *amo* genes that encode ammonia monooxygenases, and by nitrite-oxidizing bacteria (NOB) with *nap* genes that encode nitrate reductases. In contrast, denitrification, the anaerobic reduction of NO_3, NO_2, and nitric oxide (NO) to nitrous oxide (N_2O) or N_2, is the major biological mechanism by which fixed N returns to the atmosphere, thereby completing the N cycle. Nitrite reductase, which catalyzes the reduction of soluble nitrite into gaseous nitric oxide, is the key enzyme in the denitrification process and is controlled by the *nir*S and *nir*K genes that encode the cytochrome cd1 and copper nitrite reductases, respectively (Butler and Richardson 2005). The final part of the denitrification process is coded by *nor* and *nos* genes (coding nitric oxide and nitrous oxide reductases, respectively). Since every step in the N cycle is catalyzed by a different set of enzymes, many of the genes involved in this process can serve as potential functional markers for the identification of microorganisms that mediate nitrogen transformations within any given community.

The N cycle in mesic environments, where moisture is relatively predictable and sufficient, is more or less closed and tightly coupled through the production and decomposition of organic matter (Asner et al. 1997). This is because organic nutrient pools often accumulate over long time periods and most of the N required for primary production is supplied by the mineralization of stored organic matter rather than external inputs (Collins et al. 2008). However, in ecosystems that receive less than 600 mm mean annual precipitation, considered as arid and semi-arid ecosystems, pulsed water events have significant consequences on below-ground nutrient cycling through a series of soil wet–dry cycles (Austin et al. 2004). An example of this can be found in the Australian desert, where the N cycle is governed by the limitation of C and N that arises from patchy grass cover combined with fluctuations in rainfall (Cookson et al. 2006). Similarly, in the deserts of North America, patches of mesquite (*Prosopis*) and creosote bush (*Larrea*) represent islands of fertility within widespread arid soils (Schlesinger et al. 1996). This is also evident in the Atacama Desert, where the presence of any water or vegetation increases overall productivity and microbial diversity (Aguilera et al. 1999).

Given that primary production is largely constrained by water availability, N pools are more variable both temporally and spatially. Therefore, microbial community structure in desert soils is probably the result of multiple stressors on the ecosystem, such as limitation by N and low water availability (Cary et al. 2010; Gundlapally and Garcia-Pichel 2006; Nagy et al. 2005; Ríos et al. 2010). This fact is consequence of the rapid soil microbe response to incident moisture and temperature that modify the balance between N immobilization and mineralization, determining C and N turnover. Coupled with this, N inputs and losses are also tightly linked to pulses of water. For example, N fixation performed by microbial crust communities at the soil surface is generally linked to the amount of time the soil crusts are wet and thereby able to maintain activity (Belnap and Lange 2001). Even the small amount of precipitation that usually occurs in most deserts (<3 mm; Loik et al. 2004) does not elicit a response from vascular plants but can be large enough to provoke a response in crust communities. In contrast, N is lost in gaseous forms during denitrification, as well as through runoff and deep percolation. In the past, it was believed that conditions in desert ecosystems were unsuitable for denitrification, given infrequent periods of adequate water availability. However high NO flux rates have been reported in field studies that occur following wetting during the summer when temperatures are relatively high (Davidson et al. 1993; Smart et al. 1999). This is especially found in soils with low organic C because the NO consumption is lower; in consequence, there are higher rates of NO emission from soils of more arid ecosystems (Stark et al. 2002). As we can see, the gaseous losses due to denitrification can proceed at high rates during brief windows of high water, depending of nutrient availability, and can occur at rates comparable to that of temperate ecosystems.

Arid and semiarid lands have more heterogeneous distributions of resources as well as greater nutrient concentrations and microbial densities in vegetated areas and microbial crusts than in bare soils. However, the contribution of each compartment to the entire N budget in these ecosystems is poorly understood. In this chapter, we summarize the studies performed in the terrestrial part of CCB regarding different aspects of the N cycle. We present selected results from two different studies that contrast microbial diversity and specific N transformations in (i) different soil moisture conditions and (ii) different temperatures. In the first study case, we sampled two contrasting sites with different moistures, and the resulting data gave us a glimpse of CCB's soil microbial life and also highlights the need to avoid further desiccation of this unique oasis. In particular, the dry site in our investigation has lower nutrient availability, lower bacterial diversity, and lower functional diversity for the N cycle, which could make it more sensitive to water loss. The most studied N fixers in arid lands are the diazotrophs of microbial crusts, where free-living N fixers are the principal contributors to the N budget of the crust. Therefore, in the second study case, we investigated the functional (ecological) equivalence of different microbial communities within biological soil crusts by characterizing the community structure and the activity of N2-fixing microorganisms found in two arid ecosystems, Chihuahuan (CCB) and Sonoran (BC) deserts, with contrasting temperature–precipitation regimes (summer and winter rain). Community structure was determined through a culture-independent approach by fingerprinting the *nif*H gene diversity (TRFLPs). To determine N2-fixing potential, we measured nitrogenase activity by acetylene reduction assay (ARA) in two contrasting temperatures (15 and 30 °C). We found significant differences between communities in diversity, composition, and functional measures. We hypothesize that differences in nitrogenase activity between these soil crust microbial communities of contrasting habitats are driven by differences in identity of the abundant diazotrophic groups. Further research should focus on the direct measurement of activity of these groups via quantitative expression analyses.

Although microbial crusts are important components of many desert ecosystems, there is very little evidence that N fixed within these crusts actually becomes available to higher plants. In light of this, N fixers in the rhizosphere of plants become relevant. In the last part of the chapter, we compare the potential composition of microbial N fixers and denitrifier communities present in bare soils with the rhizosphere of *Agave lechuguilla*, one of the most representative plant species in the Mexican arid zones. In general, all these data suggest that environmental changes, such as moisture reduction, changes in temperature, and vegetation removal, could dramatically affect the terrestrial N cycle in CCB and stand as an example of what might occur in many arid and semiarid ecosystems that are highly sensitive to global environmental change. These data give us a glimpse of the soil microbial life of the CCB and highlight the need to avoid further desiccation of this unique oasis. Identifying abrupt changes or tipping points in the relationship between aridity and ecosystem N cycling can reveal critical vulnerabilities of dryland ecosystems to global climate change.

N Fixers and Denitrifiers in Soils with Different Water Availability

The CCB is a place where the groundwater rises to the surface through the dissolution of limestone, resulting in an "archipelago" of more than 300 pools surrounded by saline soils rich in calcium sulfates but extremely poor in nutrients. Vegetation cover and water availability in the CCB soils depend on the degree of proximity to the aquatic systems and on rainfall. This allows us to test the effects of the heterogeneous distribution of water in the same geographical region. As we can see in the rest of this book, many investigations have addressed microbial diversity in the aquatic systems of the CCB and concluded that these systems have a high microbial diversity and endemism. However, few studies have addressed CCB's soil microbial diversity. In one of the first studies of the soils, we explored the distribution and diversity of genes involved in the N cycle (*nif*H, *nir*K, and *nir*S) in two soils with different moisture levels separated by a distance of 1 km (López-Lozano et al. 2012). During February 2007, samples from the first 10 cm of top soils were collected from 8 × 8 m plots at two sites and analyzed. The first site (moist) is located on the border of a small river where the vegetation is domi-nated by the grass *Sporobolus airoides*, which covers 60% of the total plot area. The second site (dry) is situated at a distance of 250 m from a main desiccation lagoon. In contrast to the moist site, the vegetation at the desiccation lagoon site predominantly consists of the gypsophilic grass *Sesuvium erectum*, which covers around the 10% of the total plot area. Naked soil with a microbial crust consti-tuted the majority of the coverage.

The C:N:P ratios for the dry and moist sites were 104:5:1 and 296:16:1, respec-tively. Since the "Redfield ratio" in soils is 186:13:1 (Cleveland and Liptzin 2007), these ratios suggest that N is a more strongly limiting nutrient at the dry site than at the moist site. The general physicochemical data confirm that the two sites dif-fer in almost all measured parameters (Table 2.1). Soil samples from the two sites were very alkaline, but the sample from the dry site had a higher concentration of cations and anions. As expected, soil samples from the moist site had higher water content (27% soil moisture) than those from the other site (11%). The soil con-tents for almost all measured forms of C, N, and P were also higher in the moist site, with the exception of inorganic dissolved P content, which was higher at the dry site. Interestingly, nitrate was below the detection limit in the dry site.

The presence of more dense vegetation at the moist site could explain the higher concentrations of C and N. The presence of greater vegetation might also result in an increased input of organic matter into the soil, which would thus increase the organic acid content and lower the soil's pH. As a consequence of higher water availability, there is also likely to be a greater availability of organic nutrient forms, which would promote soil metabolism and lead to a more diverse but uniform microbial community. As expected from a richer community where nutrient cycling is more efficient, nutrient measurements indicate that dissolved organic C (DOC) content was four times higher at the moist site. DOC is the principal source

Table 2.1 Soil physicochemical parameters (mean ± SE) of the two studied sites located within the Chirince system in the Cuatro Cienegas Basin, Mexico

	Dry	Humid	
Soil moisture (%)	11 ± 0.6	27 ± 1.0	*
pH	8.8 ± 0.04	8.5 ± 0.04	*
Cations (cmol (+).kg^{-1})	183 ± 13	98 ± 19	*
Anions (cmol (−).kg^{-1})	16 ± 2.3	5 ± 0.6	*
Total organic C (mg.g^{-1})	3 ± 0.2	15 ± 1.5	*
Total N (mg.g^{-1})	0.11 ± 0.01	0.80 ± 0.04	*
Total P (mg.g^{-1})	0.024 ± 0.003	0.05 ± 0.003	*
Dissolved organic C (μg.g^{-1})	38 ± 5	152 ± 16	*
Dissolved organic N (μg.g^{-1})	2.1 ± 0.2	2.4 ± 0.3	*
Dissolved organic P (μg.g^{-1})	0.17 ± 0.06	0.56 ± 0.20	*
Inorganic dissolved P (μg.g^{-1})	0.012 ± 0.02	0.003 ± 0.001	*
Ammonium (μg.g^{-1})	0.43 ± 0.03	0.68 ± 0.06	*
Nitrate (μg.g^{-1})	0.0 ± 0.0	0.24 ± 0.13	ns
Microbial C (μg.g^{-1})	107 ± 17	595 ± 133	*
Microbial N (μg.g^{-1})	5.3 ± 1.8	11.0 ± 2.3	*

*Significant difference among sites $p < 0.05$; *ns* no significant

of C for heterotrophic microorganisms as observed in the moist site; this could promote higher heterotrophic microbial activity (Montaño et al. 2007). In contrast, when the availability of DOC is low, autotrophic pathways, such as nitrification, would be favored, and nitrate would accumulate (Vitousek 2002). Nevertheless, even though the organic carbon was so different between sites, we found low concentrations of nitrate at both sites (undetectable in the dry site). This could be attributed to low temperatures (0–10 °C) at the time of sampling, which may have inhibited soil metabolism and concomitantly lowered rates of denitrification. In other deserts, surface concentrations of nitrate have been found to be low (near zero or undetectable), but large quantities are sequestered below a depth of 1 m because not all the available nitrate is consumed in the soil zone or returned to the atmosphere (Walvoord et al. 2003).

To explore the distribution and diversity of functional groups involved in the N cycle, we constructed clone libraries of *nif*H, *nir*K, and *nir*S genes by direct DNA extraction from soil samples. One hundred partial sequences of *nif*H were obtained for each site, which belonged to 25 different OTUs (assigned at 97% of identity) (Fig. 2.1a). A total of 19 OTUs were present in the humid site, whereas only 9 OTUs were present in the dry site and only 3 were present in both sites (shared between sites). Out of the total OTUs identified, 12 were associated to Deltaproteobacteria, 7 to Alphaproteobacteria, 3 to Gammaproteobacteria, and 3 to Cyanobacteria. The most abundant OTU present at both sites was related to Rhizobiales. This observation, together with the absence of Leguminosae in both sites, suggests the presence of nonsymbiotic *Rhizobiales* in the diazotrophic soil community. The different kinds of phototrophs and potential N fixers observed in the data (besides Cyanobacteria) include ancient taxa such as *Chloroflexi* and the *Alphaproteobacteria Porphyrobacter*. The latter is a genus that can fix C under aerobic conditions in the

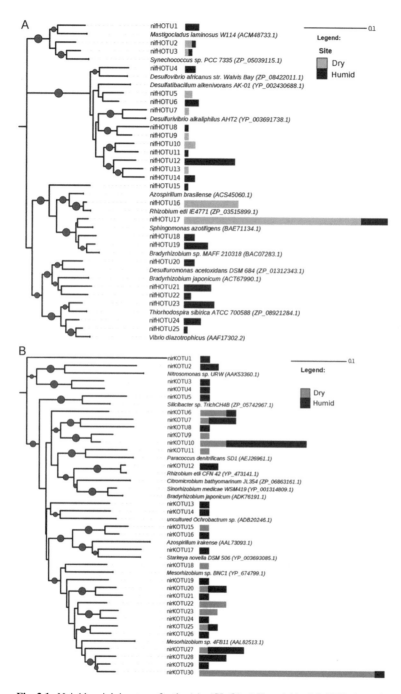

Fig. 2.1 Neighbor-joining trees for the (**a**) *nif*H, (**b**) *nir*K, and (**c**) *nir*S OTUs from the two studied sites: dry and humid, within the Churince system in CCB, Mexico. Gray circles indicate nodes with bootstrap support >0.5. Relative abundances are shown as gray bars for the dry site and as black bars for the humid site

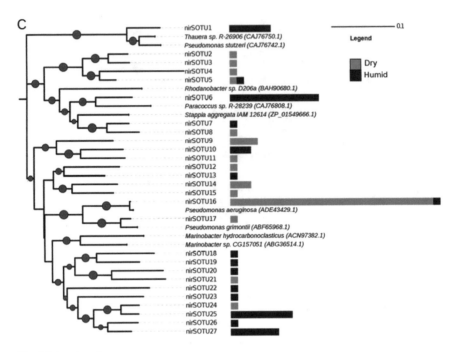

Fig. 2.1 (continued)

dark (Hiraishi and Imhoff 2005), while the class Chloroflexi contains photosynthetic, anoxygenic, nonsulfur bacteria typically found in microbial mats in salty, shallow marine environments (Hanada and Pierson 2006). We obtained many undescribed *nif*H sequences from a relatively small sample, confirming the uniqueness of the CCB soils. On the other hand, the dominance of *Pseudomonas pseudoalcaligenes, Azospirillum brasilense,* and *Rhizobium* sp. has been reported in association with roots of the drought-tolerant grass *Lasiurus sindicus* in the Thar Desert of Rajasthan, India (Chowdhury et al. 2009). This observation suggests that the grass species found in the CCB could be the source of the Rhizobiales and their associated nitrogenases. However, no symbiosis between halotolerant grasses and Rhizobiales has yet been reported.

A total of 50 sequences of *nir*K were obtained for each site, from which 30 OTUs could be identified (assigned at 97% of identity) (Fig. 2.1b). The majority of the sequences could not be associated with known sequences in public databases. A total of 24 and 13 OTUs were present in the moist and dry sites, respectively. From these, only seven OTUs were shared between sites. One single OTU could be associated with *Nitrosomonas* sp., a common group associated with denitrification processes, and another single OTU with *Rhizobium etli*, whereas the rest of the OTUs are highly divergent or have a high similarity with undescribed taxa. For *nir*S, 50 sequences were obtained for each site. A total of 27 OTUs were recovered (assigned at 95% of identity), 15 from the moist site, 14 from the dry site, and only 2 shared

OTUs (Fig. 2.1c). As in the case of *nir*K, most *nir*S sequences could not be associated with any known sequence reported in public databases. The only sequence with a database hit was related to *Pseudomonas*. This OTU was present at both sites, being abundant at the dry site but rare at the moist site. The other shared OTU was rare in both sites. The apparent uniqueness of the CCB denitrifiers may be the result of the low global sampling or an artifact of fewer deposited sequences from isolated strains. However, this small sample suggests that a highly diverse denitrifying community is waiting to be discovered in CCB soils. In particular, the dry site had a lower nutrient availability and lower functional diversity for the N cycle, which could make it more sensitive to water loss.

N Fixers in Soils with Different Precipitation Regimes: The Case of Soil Crusts of Cold and Hot Deserts

So far, we have presented evidence for great variation in the presence and abundance of N cycle genes under different regimes of soil moisture in desert soils of CCB. Despite this evidence, it is commonly assumed that changes in microbial composition of communities are not relevant given the functional redundancy found in different microbial groups. In other words, it does not matter whether the phylogenetic or geographic origin of the groups are present, as long as "the genes" are present to perform the function of interest. However, few studies have explicitly tested this assumption. The assumption of functional equivalence implies that local adaptations do not exist and that even rates of specific transformations will be equivalent or equal as long as the functional genes are present. In this section, we summarize investigations addressing the functional (ecological) equivalence assumption of different microbial communities found in biological soil crusts (BSC) by characterizing the community structure and activity of N_2-fixing microorganisms from two arid ecosystems: CCB and Valle de Guadalupe in Baja California (BC). These two regions have contrasting temperature–precipitation regimes of summer and winter rain. We hypothesize that, given the historical differences in precipitation regimes, not only will BSC community structures differ but also their functional rates of potential N_2 fixation. To test this hypothesis, we conducted a common garden experiment for both communities under two different constant temperatures (15 and 30 °C) and constant humidity (90–100% soil saturation). Community structure was determined through a culture-independent approach by fingerprinting the *nif*H gene diversity (TRFLPs). To determine N_2-fixation potential, we measured nitrogenase activity by acetylene reduction assay (ARA) and used a two-way ANOVA to test for the effects of sample origin (a proxy for microbial composition), the incubating environment, and their interaction on functional response (Table 2.2).

We found significant differences between communities in diversity, composition, and functional measures. Results from the functional assay indicated that crust origin (a proxy for composition) (Fig. 2.2) had a significant effect on nitrogenase activity, which implies that no functional equivalence can be assumed simply by the

Table 2.2 Two way ANOVA test for the effect of composition and incubation temperature N₂-fixation potential

	Sampling time	Origin (proxy for composition)	Temperature	Origin × temperature
N₂-fixation potential	10 days	$F_1 = 14.8\ p < 0.001$	$F_1 = 16.8\ p < 0.001$	$F_1 = 0.75\ p = 0.8$

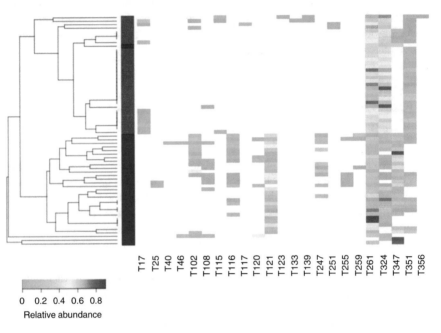

0   0.2 0.4 0.6 0.8
Relative abundance

Fig. 2.2 UPGMA cluster analysis (left) and heatmap (right) of the microbial composition of BSC (biological soil crust) for two sites, CCB (Cuatro Cienegas, Coahuila) and BC (Valle de Guadalupe, Baja California). UPGMA cluster is presented using a similarity index of Bray–Curtis, blue corresponds with BC samples, and red corresponds with CCC samples

presence of the marker genes of such metabolic functions. Finally, our results also serve as evidence for local adaptation of microbial communities, which has a lasting impact on the functional response of microbial communities.

Influence of the Rhizosphere on the Abundance of Microorganisms Involved in the N Cycle

To conclude this chapter, we highlight the importance of rhizosphere microorganisms (all the microbes living in the influence zone of plant roots) in nutrient cycling in CCB. As we mentioned previously, soil from vegetated areas differs from bulk soil in many aspects. Arid and semiarid ecosystems provide many sources of abiotic stress, especially in open spaces, where microbes must withstand high temperatures, desiccation, intense radiation, and nutrient scarcity. Abiotic stress is partly alleviated within the plant patches, which filter the quantity and quality of light and allow accumulation of water and resources by the excretion of root exudates (Bochet et al. 1999; Goberna et al. 2007). Soil bacterial communities underneath plant patches have a higher biomass, are more metabolically active, and show higher microbial respiration rates (Goberna et al. 2007). Therefore, these arid ecosystems can be viewed as mosaics, where low productivity and high-productivity habitats combine to create systems with nutrient cycle steps that can occur in different proportions.

To examine this, we compared the potential composition of microbes involved in important nutrient processes present in bare soils and in the rhizosphere of *Agave lechuguilla*, one of the most representative plant species in the Mexican arid zones and a dominant plant in the xeric shrub of CCC. In this work, we sampled rhizosphere communities from *A. lechuguilla* (vegetation patch) and from the first 10 cm of vegetation-free bulk soil (gaps) at four sites of CCC. The 16S rRNA gene was amplified and sequenced by Illumina to characterize the bacterial community (24 samples from vegetation plots and 24 from gaps). We categorized the functional traits of the bacteria classified at the genus level with confidence thresholds equal or over 80% according to the Naïve Bayesian Classifier (Wang et al. 2007). Relative abundances of each category were calculated as the proportion of sequences of each OTU in either patch or gap. Candidate divisions, for which no trait information is available in the literature, constituted a small fraction of the total community in both patches and gaps. We specifically considered eight traits that confer the ability to obtain three potentially limiting resources: carbon, nitrogen, and phosphorus. We reviewed the literature to look for:

1. C fixation that allows aerobic or anaerobic growth using CO_2 as a sole carbon source, with either facultative or obligate detected under laboratory conditions
2. N fixation due to either the possession of *nif* genes or detected under laboratory conditions
3. Ammonia oxidation to hydroxylamine, either due to the possession of *amo* genes or detected under laboratory conditions
4. Nitrate reduction to nitrite, which is the first step of denitrification but is not unique to denitrifying organisms, either due to the direct detection of *nap* genes or the metabolic effect detected under laboratory conditions
5. Denitrification with involvement in any step from the reduction of nitrite to the production of molecular nitrogen, due to the possession of *nir*, *nor*, and/or *nos* genes or detected under laboratory conditions
6. P solubilizing by the presence of phosphatase genes (phospho-mono-esterase and phosphor-diesterase) or detection under laboratory conditions

High-throughput sequencing allows for a deeper exploration of microbial diversity but trait definition is constrained to the closest cultured representative of each query sequence. However, trait assignment to closely related bacteria is supported by the widespread phylogenetic conservatism of many functional traits in

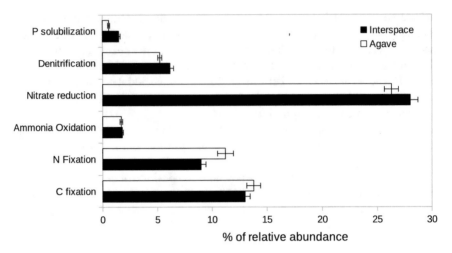

Fig. 2.3 Percentage of relative abundance of functional guilds compared between vegetation patches (Agave) and soil gaps (interspace), obtained from samplings of rhizosphere communities from *A. lechuguilla* and from the first 10 cm of vegetation-free bulk soil at four sites of CCB

prokaryotes (Martiny et al. 2013). Similar approaches based on trait assignment to close relatives have been recently proposed for the prediction of gene contents (Kembel et al. 2012; Langille et al. 2013; Goberna et al. 2014).

Results of our analysis found a higher relative abundance of bacteria carrying the *nif*H gene in the rhizosphere of *A. lechuguilla* than found in the gaps (Fig. 2.3), represented principally by Sphingomonadales and Rhizobiales. Many ecosystems lack large numbers of plants that are in symbiosis with N fixers and, therefore, likely depend upon free-living N fixers that are associated with the rhizosphere of nonsymbiotic plants to provide a source of new N to those ecosystems (Reed et al. 2011). As expected, N fixers were less abundant in gaps. It is probable that lower organic matter in the gaps acts as a factor limiting the reducing power needed to fix N since N fixation is a high, energy-demanding process. Other functional groups that showed differences in relative abundance between gaps and patches were denitrifiers and P solubilizers. These were both more abundant in the gaps (Fig. 2.3). Specifically, the denitrifiers were primarily represented by the genera *Streptomyces, Euzebya, Bacillus,* and *Bosea.* Multiple factors have been shown to influence the abundance and activity of denitrifiers, including soil texture (D'Haene et al. 2003; Gu and Riley 2010; Gu et al. 2013), pH (Baggs et al. 2010; Čuhel et al. 2010; Ligi et al. 2014; Mørkved et al. 2007), organic material (OM) (Barrett et al. 2016; Mosier et al. 2002), and the amount of inorganic N (Niboyet et al. 2009). In general, our soil samples did not present significant differences in texture or pH but had higher concentrations of OM and nitrate in the patches. It is therefore possible that a strong competition for nitrate occurs between plants and denitrifiers in patches, resulting in the higher abundance of denitrifiers observed in the gaps. P solubilizers were represented by members of the family Gaiellaceae, a group in which alkaline and acid phosphatase activities have been reported (Albuquerque and da Costa 2014). However only two unknown genera were found. For this reason, we cannot differentiate whether their presence is due to habitat preference or to the functional guild to which they belong.

It is clear that, in patches, plants release significant amounts of photosynthetic C into the rhizosphere through root exudation. Root exudates are considered the principal suppliers of energy supporting enhanced microbial abundance and activity in the rhizosphere compared to bulk soils (Kapoor and Mukerji 2006; Coskun et al. 2017). Thus, plants are able to modify their environmental conditions in the root vicinity, a process termed "mining strategy" (Erel et al. 2017; Neumann et al. 2014; Tückmantel et al. 2017). These modifications around the roots allow the plants to directly select specific microbial communities. Indeed, the carbon-rich conditions occurring in the root vicinity allow for the selection of effective nutrient-mobilizing microbial communities (Fitzpatrick et al. 2018; Nicolitch et al. 2017) that differ from the bulk soil. There is little doubt that soil properties as well as plant species influence the structure and function of microbial communities. An improved knowledge of specific plant–microorganism interactions in the rhizosphere in CCB could also be important for conservation and reforestation projects in which vegetated areas that have been depleted should be replanted with native plants. Furthermore, in this context, the impact of climate change on plant–microorganism interactions will be important to understand for such conservation and restoration efforts to be successful.

In a more recent survey of many soil sites in this gypsum-based ecosystem, a rapid degradation of the ecosystem has been observed due to accelerated water loss due to the overexploitation of the aquifer. Given the extraordinary diversity and divergence of the micro- and macrobiota of the Cuatro Cienegas valley, this degradation alerts us to the loss of invaluable genetic resources and diversity. The data presented in this chapter suggest that environmental changes such as loss of soil moisture, changes in temperature, and vegetation removal could all dramatically affect the terrestrial N cycle in CCB and endanger undescribed microbial diversity that has enormous untapped potential for agriculture and other applications.

References

Aguilera LE, Gutiérrez JR, Meserve PL (1999) Variation in soil micro-organisms and nutrients underneath and outside the canopy of *Adesmia bedwellii* (Papilionaceae) shrubs in arid coastal Chile following drought and above average rainfall. J Arid Environ 42:61–70

Albuquerque L, da Costa MS (2014) The family *Gaiellaceae*. In: Rosenberg E, DeLong EF, Lory S, Stackebrandt E, Thompson F (eds) The prokaryotes. Springer, Berlin, Heidelberg, pp 357–360

Asner GP, Seastedt TR, Townsend AR (1997) The decoupling of terrestrial carbon and nitrogen cycles. Bioscience 47:226–234

Austin AT, Yahdjian L, Stark JM, Belnap J et al (2004) Water pulses and biogeochemical cycles in arid and semiarid ecosystems. Oecologia 141:221–235

Baggs EM, Smales CL, Bateman EJ (2010) Changing pH shifts the microbial source as well as the magnitude of N_2O emission from soil. Biol Fertil Soils 46:793–805

Barrett M, Khalil MI, Jahangir MMR et al (2016) Carbon amendment and soil depth affect the distribution and abundance of denitrifiers in agricultural soils. Environ Sci Pollut Res 23:7899–7910

Belnap J, Lange OL (2001) Biological soil crusts: structure, function and management. Ecological studies, vol 150. Springer-Verlag, Berlin-Heidelberg

Bochet E, Bochet E, Rubio JL et al (1999) Modified topsoil islands within patchy Mediterranean vegetation in SE Spain. Catena 38:23–44

Butler CS, Richardson DJ (2005) The emerging molecular structure of the nitrogen cycle: an introduction to the proceedings of the 10th annual N-cycle meeting. Biochem Soc Trans 1:113–118

Cary SC, McDonald IR, Barrett JE, Cowan DA (2010) On the rocks: the microbiology of Antarctic Dry Valley soils. Nat Rev Microbiol 8:129–138

Chowdhury SP, Schmid M, Hartmann A, Tripathi AK (2009) Diversity of 16S-rRNA and *nif*H genes derived from rhizosphere soil and roots of an endemic drought tolerant grass, *Lasiurus sindicus*. Eur J Soil Biol 45:114–122

Cleveland CC, Liptzin D (2007) C:N:P stoichiometry in soil: is there a "Redfield ratio" for the microbial biomass? Biogeochemistry 85:235–252

Collins SL, Sinsbaugh RL, Crenshaw C et al (2008) Pulse dynamics and microbial processes in arid-land ecosystems. J Ecol 96:413–420

Cookson WR, Müller C, O'Brien PA et al (2006) Nitrogen dynamics in an Australian semiarid grassland soil. Ecology 87:2047–2057

Coskun D, Britto DT, Shi W, Kronzucker HJ (2017) How plant root exudates shape the nitrogen cycle. Trends Plant Sci 22:661–673

Čuhel J, Šimek M, Laughlin RJ et al (2010) Insights into the effect of soil pH on N_2O and N_2 emissions and denitrifier community size and activity. Appl Environ Microbiol 76:1870–1878

D'Haene K, Moreels E, De Neve S et al (2003) Soil properties influencing the denitrification potential of Flemish agricultural soils. Biol Fertil Soils 38:358–366

Davidson EA, Matson PA, Vitousek PM et al (1993) Processes regulating soil emissions of NO and N_2O in a seasonally dry tropical forest. Ecology 74:130–139

Devol AH (1991) Direct measurement of nitrogen gas fluxes from continental shelf sediments. Nature 349:319–321

Erel R, Bérard A, Capowiez L et al (2017) Soil type determines how root and rhizosphere traits relate to phosphorus acquisition in field-grown maize genotypes. Plant Soil 412:115–132

Fitzpatrick CR, Copeland J, Wang PW et al (2018) Assembly and ecological function of the root microbiome across angiosperm plant species. Proc Natl Acad Sci 115:*201717617*

Goberna M, Pascual JA, García C et al (2007) Do plant clumps constitute microbial hotspots in semi-arid Mediterranean patchy landscapes? Soil Biol Biochem 39:1047–1054

Goberna M, Navarro-Cano JA, Valiente-Banuet A et al (2014) Abiotic stress tolerance and competition-related traits underlie phylogenetic clustering in soil bacterial communities. Ecol Lett 17:1191–1201

Gu C, Riley WJ (2010) Combined effects of short term rainfall patterns and soil texture on soil nitrogen cycling: a modeling analysis. J Contam Hydrol 112:141–154

Gu J, Nicoullaud B, Rochette P et al (2013) A regional experiment suggests that soil texture is a major control of N_2O emissions from tile- drained winter wheat fields during the fertilization period. Soil Biol Biochem 60:134–141

Gundlapally SR, Garcia-Pichel F (2006) The community and phylogenetic diversity of biological soil crusts in the Colorado Plateau studied by molecular fingerprinting and intensive cultivation. Microbial Ecol 52(2):345–357

Hanada S, Pierson BK (2006) The family Chloroflexaceae. In: Dworkin M, Falkow S, Rosenberg E et al (eds) The prokaryotes: a handbook on the biology of bacteria. Springer Science+Business Media, New York, pp 815–842

Hiraishi A, Imhoff JF (2005) Genus Porphyrobacter. In: Brenner DJ, Krieg NR, Staley JT (eds) Bergey's manual of systematic bacteriology, the alpha-, beta-, delta- and epsilonproteobacteria. Springer Science and Business Media Inc., New York, pp 275–279

Kapoor R, Mukerji KG (2006) Rhizosphere microbial community dynamics. In: Mukerji KG, Manoharachary C, Singh J (eds) Microbial activity in the rhizosphere. Springer, Berlin, pp 55–66

Kembel SW, Wu M, Eisen JA et al (2012) Incorporating 16S gene copy number information improves estimates of microbial diversity and abundance. PLoS Comput Biol 8:e1002743

Langille MGI, Zaneveld J, Caporaso JG et al (2013) Predictive functional profiling of microbial communities using 16S rRNA marker gene sequences. Nat Biotechnol 31:814–823

Ligi LT, Truu M, Truu J et al (2014) Effects of soil chemical characteristics and water regime on denitrification genes (nirS, nirK, and nosZ) abundances in a created riverine wetland complex. Ecol Eng 72:47–55

Loik ME, Breshears DD, Lauenroth WK et al (2004) A multi-scale perspective of water pulses in dryland ecosystems: climatology and ecohydrology of the western USA. Oecologia 141:269–281

López-Lozano NE, Eguiarte LE, Bonilla-Rosso G et al (2012) Bacterial communities and the nitrogen cycle in the gypsum soils of Cuatro Ciénegas Basin, Coahuila: a Mars analogue. Astrobiology 12:699–709

Martiny AC, Treseder K, Pusch G (2013) Phylogenetic conservatism of functional traits in microorganisms. ISME J 7:830–838

Montaño NM, García-Oliva F, Jaramillo VJ (2007) Dissolved organic carbon affects soil microbial activity and nitrogen dynamics in a Mexican tropical deciduous forest. Plant Soil 295:265–277

Mørkved PT, Dörsch P, Bakken LR (2007) The N_2O product ratio of nitrification and its dependence on long-term changes in soil pH. Soil Biol Biochem 39:2048–2057

Mosier AR, Doran JW, Freney JR (2002) Managing soil denitrification. J Soil Water Conserv 57:505–512

Nagy ML, Pérez A, Garcia-Pichel F (2005) The prokaryotic diversity of biological soil crusts in the Sonoran Desert (Organ Pipe Cactus National Monument, AZ). FEMS Microbiol Ecol 54:233–245

Neumann G, Bott S, Ohler M et al (2014) Root exudation and root development of lettuce (Lactuca sativa L. cv. Tizian) as affected by different soils. Front Microbiol 5(1–6):2

Niboyet A, Barthes L, Hungate BA et al (2009) Responses of soil nitrogen cycling to the interactive effects of elevated CO_2 and inorganic N supply. Plant Soil 27:35–47

Nicolitch O, Colin Y, Turpault MP et al (2017) Tree roots select specific bacterial communities in the subsurface critical zone. Soil Biol Biochem 115:109–123

Reed SC, Cleveland CC, Townsend AR (2011) Functional ecology of free-living nitrogen fixation: a contemporary perspective. Annu Rev Ecol Evol Syst 42:489–512

Ríos A, Valea S, Ascaso C et al (2010) Comparative analysis of the microbial communities inhabiting halite evaporites of the Atacama Desert. Int Microbiol 13:79–89

Schlesinger WH, Raikes JA, Hartley AE et al (1996) On the spatial pattern of soil nutrients in desert ecosystems. Ecology 77:364–374

Smart DR, Stark JM, Diego V (1999) Resource limitations to nitric oxide emissions from a sagebrush-steppe ecosystem. Biogeochemistry 47:63–86

Stark JM, Smart DR, Hart SC et al (2002) Regulation of nitric oxide emissions from forest and rangeland soils of western North America. Ecology 83:2278–2292

Tückmantel T, Leuschner C, Preusser S et al (2017) Root exudation patterns in a beech forest: dependence on soil depth, root morphology, and environment. Soil Biol Biochem 107:188–197

Vitousek PM (2002) Nutrient cycling and limitation. Hawaii as a model system. Princeton University Press, Princeton, NJ

Walvoord MA, Phillips FM, Stonestrom DA et al (2003) A reservoir of nitrate beneath desert soils. Science 302:1021–1024

Wang Q, Garrity GM, Tiedje JM et al (2007) Naive Bayesian classifier for rapid assignment of rRNA sequences into the new bacterial taxonomy. Appl Environ Microbiol 73:5261–5267

Zehr JP, Kudela RM (2011) Nitrogen cycle of the open ocean: from genes to ecosystems. Annu Rev Mar Sci 3:197–225

Chapter 3
The Effect of Nutrients and N:P Ratio on Microbial Communities: Testing the Growth Rate Hypothesis and Its Extensions in Lagunita Pond (Churince)

James Elser, Jordan Okie, Zarraz Lee, and Valeria Souza

Contents

Abstract The absolute and relative supplies of nutrients such as nitrogen and phosphorus can affect ecosystem properties and microbial biodiversity. More recently, the theory of biological stoichiometry has advanced connections between ecosystem ecology and cellular/molecular biology by proposing a link between biochemical features of microbial cells and their nutrient ratios. Specifically, the growth rate hypothesis (GRH) postulates that cellular stoichiometry varies according to growth rate due to increased allocation to P-rich ribosomal RNA to support rapid growth. Expanding on the GRH, it is predicted that microbes have a suite of genomic features that determine the ability to achieve rapid growth and, hence, influence biomass N:P. These genomic features include codon usage bias, number of rRNA and tRNA genes, and genome size, all of which have been individually linked to growth rate and fitness. This chapter discusses two experiments conducted at the Churince

J. Elser (✉)
School of Life Sciences, Arizona State University, Tempe, AZ, USA

Flathead Lake Biological Station, University of Montana, Polson, MT, USA
e-mail: Jim.Elser@umontana.edu

J. Okie · Z. Lee
School of Life Sciences, Arizona State University, Tempe, AZ, USA

V. Souza
Departamento de Ecología Evolutiva, Instituto de Ecología, Universidad Nacional Autónoma de México, Coyoacan, Mexico

© Springer International Publishing AG, part of Springer Nature 2018   31
F. García-Oliva et al. (eds.), *Ecosystem Ecology and Geochemistry of Cuatro Cienegas*, Cuatro Ciénegas Basin: An Endangered Hyperdiverse Oasis,
https://doi.org/10.1007/978-3-319-95855-2_3

system in Lagunita pond to test the GRH. Churince is an ideal location for this because surface waters in this region have highly imbalanced N:P stoichiometry (TN:TP atomic ratio >100), where P is likely to be strongly limiting. The first experiment was a replicated in situ mesocosm experiment comparing three different nutrient treatments with varying N:P to a control treatment. The second experiment was a whole-pond perturbation, fertilizing with nutrients of N:P = 16 and using metagenomics to compare responses to replicated, internal control mesocosms. We discuss changes in microbial biomass N:P, species composition, and genomic features of the microbes in response to these perturbations of nutrient supplies and N:P.

Keywords Churince system · Growth rate hypothesis · Phosphorus · Nutrient stoichiometric ratios · Ribosomal RNA

Introduction

The famous clarity of the pozas of Cuatro Ciénegas is a direct manifestation of its low-nutrient conditions ("oligotrophy"). That is, very low concentrations of key nutrient elements (e.g., nitrogen (N) and especially phosphorus (P)) constrain the proliferation of microbes, algae, and aquatic vegetation. N and P limitation of ecosystem production is widespread globally (Elser et al. 2007), and, indeed, overcoming the low N and P status of soils via fertilizer application was essential in the "Green Revolution" that greatly enhanced agricultural production during the last century. Furthermore, not only are nutrient concentrations at CCB low, they are also imbalanced. The overall ratios of N and P are very high (often 75:1 or higher, by moles here and throughout), considerably higher than those seen, for example, in the world's oceans (16) or greater than those thought to drive ecosystem P limitation (~28–30; Downing and McCauley 1992). So, in most of CCB's ecosystems, we expect strong P limitation to be operating both in ecological as well as evolutionary time. While the effects of N and P limitation on both aquatic and terrestrial systems are very well-studied in various habitats, their impacts on the evolution of living things are less well-investigated. Some evidence of impacts of nutrient limitation on evolutionary processes can be seen in a variety of studies and in particular in patterns of biomass C:N:P ratios in various biota. For example, there are phylogenetic differences in the C:N:P ratios of terrestrial insects (Woods et al. 2004; Fagan et al. 2002), poleward trends in P content (C:P ratio) in congeneric taxa of *Daphnia* (Elser et al. 2000a), and even macroevolutionary patterns in the investment of elements (C, N, and S) in the amino acids making up microbial proteomes (Elser et al. 2011). However, evolutionary studies within a stoichiometric framework are just in their infancy. The extreme stoichiometric imbalance and pronounced P limitation at CCB provide an excellent opportunity to test impacts of P limitation on genomic evolution. But what expectations can be developed for the evolutionary underpinnings of stoichiometric variation in biota?

To explain observed variation in the N:P ratios of various biota (such as low N:P *Daphnia* vs high N:P copepods) and thus, potentially, their susceptibility and responses to N or P limitation, the growth rate hypothesis (GRH) was proposed (Elser et al. 2000b). In GRH, low N:P ratios in living things are attributed to increased overall allocation (in terms of percent of dry mass) to P-rich ribosomal RNA needed to meet the protein synthesis demands of fast growth rate. The hypothesis was developed to apply primarily to small-bodied heterotrophic organisms, excluding large-bodied animals due to the importance of P-rich bone in vertebrates (Elser et al. 1996) and excluding plants and other phototrophs due to their considerable capacity for luxury uptake and storage of elements such as P. A large number of experimental and observational tests of the GRH has emerged subsequently, generally lending support for the GRH in both intraspecific (physiological, ontogenetic) and interspecific comparisons. More specifically, Hessen et al. (2013) completed a meta-analysis of more than 70 published experiments testing the GRH in various invertebrate taxa and in bacteria. In these studies, authors reported statistically significant ($p < 0.05$) positive correlations between growth rate and RNA content, growth rate and P content, and RNA content and P content in 67–71% of the individual assessments. Thus, we can conclude that the GRH has been established as a sound foundation for understanding variation in biomass C:N:P ratios in microbes and small consumers, critical heterotrophic components of most ecosystems. But what is the genomic basis of the GRH and how has it evolved?

With recent increases in genomic information, the GRH has expanded to include a set of genomic characteristics that support growth rate. Elser et al. (2000b) hypothesized that one key genetic component contributing to production and support of elevated ribosome content is the number of copies of the ribosomal RNA (rRNA) operon in the genome. This is because rates of protein synthesis can be limited by the number of RNA polymerase molecules transcribing the rRNA gene. This limitation is overcome by increasing the rRNA gene dosage in the genome and the strategic placement of the gene(s) near the origin of replication (Wagner 1994). Support for this idea comes from various lab studies. For example, in *E. coli*, which typically has seven copies of rRNA genes in its genome, a prolonged lag time in response to nutrient availability was observed when three or four copies of the rRNA gene are deleted (Condon et al. 1995). High rRNA gene copy number is required for *Bacillus subtilis* to successfully grow after germination of spores (Yano et al. 2013). Exploring the growth rate of 184 strains of environmental isolates, Roller et al. (2016) identified a positive relationship between maximum growth rate and rRNA gene copy number. Furthermore, copiotrophic bacteria typically found in feast and famine condition or early successional species tend to have higher rRNA gene copy number than their phylogenetic counterparts in oligotrophic environments (Klappenbach et al. 2000; Lauro et al. 2009; Nemergut et al. 2016). However, increasing the amount of ribosomes (the enzyme) can only influence protein synthesis rates if it is accompanied by increased availability of the substrate, tRNA. To support the tRNA requirement, the abundance of tRNA genes in the genome is positively correlated with the rRNA gene copy number (Lee et al. 2009). Maintaining a high density of ribosomes and associated tRNA molecules may not on its own be sufficient to maintain maximally efficient protein synthesis if the genome codes for amino acids that are not well-matched to the organism's own genomic tRNA pool. Thus, there can be selection for use of amino acids that achieve faster and more accurate translation rate, especially in highly expressed genes such as those that code for ribosomal proteins (Rocha 2004).

The genomic capabilities to support high growth rate just described come with a cost. For example, it is hypothesized that bacteria with increased rRNA gene copy number are less efficient at resource utilization and have a higher basal nutrient requirement. In *E. coli*, the rRNA gene P2 promoter is a constitutive promoter that is weakly regulated and maintains a low basal level of rRNA synthesis (Maeda et al. 2016; Paul et al. 2004). If a similar promoter structure is present in other bacteria, we can expect bacteria with higher rRNA gene copy number to have a higher basal level of rRNA synthesis and potentially higher cellular ribosome content and P requirement. The higher basal level is important for rapid response but will be more costly to maintain in a low-nutrient environment. The higher gene dosage also means that the cell needs to maintain a larger genome size and thus will have a higher P demand. A positive correlation between rRNA gene copy number and genome size exists for eukaryotes, although the underlying mechanisms are still unclear (Prokopowich et al. 2003). A more recent analysis of eukaryotic genomes also found larger genomes to have a higher proportion of duplicated genes than smaller genomes (Elliott and Gregory 2015). Similarly, bacteria with fewer rRNA genes tend to have smaller genomes while bacteria with higher rRNA gene copy number have larger genomes that also carry more biosynthetic pathways (Roller et al. 2016). Therefore, high growth rate capacity may become a burden to the cell in environments with chronically low-nutrient conditions.

Vieira-Silva and Rocha (2010) have synthesized these genomic traits as components of the "systemic imprint" of growth, allowing us to expand the GRH to include a broader genomic component. In this synthetic view, microbes with high rRNA and tRNA gene copy number have a higher capacity for ribosome production to support increased protein synthesis demands of rapid growth. Furthermore, there will be selection for efficient translation in such fast-growing taxa, leading to increased codon usage bias. Finally, carrying increased numbers of genes for ribosome production itself can contribute to increase in genome size. Thus, increased pools of ribosomes and tRNA molecules along with larger genomes all require investment of P-rich nucleic acids, which will be manifested as low biomass C:P and N:P ratios of such taxa and will imply that these taxa will perform poorly in low-nutrient, low N:P environments. These combined genomic and physiological adaptations have been investigated primarily at the organismal level (Gyorfy et al. 2015; Fu and Gong 2017). Furthermore, a recent study by Hartman et al. (2017) demonstrated a support for this hypothesis, showing that soil microbes associated with increased carbon turnover (faster growth) had higher rRNA gene copy number and that this relationship is constrained by inorganic P availability.

Over the course of two summers (2011 and 2012), we performed experiments to test the GRH and the genomic signature concept at the community level. The results we synthesize here are reported in Lee et al. (2015, 2017) and Okie et al. (in review). Specifically, we tested how these genomic signatures change as community structure shifts in response to changes in nutrient supply and N:P stoichiometry. We expect the microbial community in unenriched conditions to have a genomic signature indicating dominance of species with a low growth rate lifestyle. However, nutrient enrichment will induce a shift to microbes that have genomic signatures indicative of high growth rate and low biomass C:P and N:P ratios.

Experiments/Tests

Study site: We tested these ideas in a pair of field experiments in Lagunita Pond (latitude: 26° 50′ 53.19″ N, longitude: 102° 8′ 29.98″ W). Lagunita is a shallow pond (~25 cm maximum depth) located adjacent to a larger lagoon (Laguna Intermedia) in the western region of the Cuatro Ciénegas Basin (CCB; Fig. 3.1a). Lee et al. (2015) report detailed physical and chemical characteristics of the pond. Briefly, Lagunita's waters are high in conductivity and dominated by Ca^{2+}, SO_4^{2-}, and CO_3^{2-}, while phosphorus (P) concentrations are moderate (total P: 1.79 µmol/L) but highly stoichiometrically imbalanced with nitrogen (N) with average total N:P ratios of ~122 (molar). The pond experiences strong evaporation during summer, and thus, in both experiments, water depths declined significantly during the study.

Methods

Experiment 1: Experiment 1 sought to test the GRH by manipulating P availability at different N:P ratios in small, replicate enclosures that were accessed by constructing a welded steel superstructure above the pond (Fig. 3.1b). Enclosures consisted of clear cylinders (40 cm diameter) inserted several centimeters into the sediments (Fig. 3.1c and d). There were four treatments: unenriched (no added nutrients), +P (P only added), +NP16 (N and P added at a 16:1 molar ratio), and +NP75 (N and P added at a 75:1 molar ratio). Each treatment was replicated four times, and the enclosures distributed in Lagunita according to a randomized block design. In each of the enriched treatments, P was added (as KH_2PO_4) every 3–4 days to achieve a target concentration of 1 µM after PO_4 concentrations in each enclosure were checked by colorimetric analysis of soluble reactive P (SRP). In the NP16 and NP75 treatments, N was added (as NH_4NO_3) with P to achieve the target ratio. Nutrient enrichment was sustained until day 21, and the experiment ran for 42 days in total. Basic limnological variables were monitored, and on days 6, 21, and 42, measurements of chlorophyll biomass in the water column were made. On days 21 and 42, samples were collected for determination of microbial cell densities, major nutrient concentrations, and seston biomass and C:N:P ratios. For analysis of microbial communities, samples of both surficial sediments and of the water column were obtained and flash frozen for later analysis of 16S and 18S profiling via 454 sequencing after PCR amplification. Finally, water column DNA samples were subjected to qPCR analysis for assessment of 16S rRNA gene copy numbers; these data were later normalized to cell abundance from cell counts.

Fig. 3.1 (**a**) Lagunita is a small pond located to the west of Laguna Intermedia in the Churince drainage. (**b**) Construction of the steel scaffold that allowed Lagunita mesocosms to be sampled without disturbing the sediments. (**c**) Administering the nutrient enrichment treatments in Experiment 1. (**d**) An experimental block from Experiment 1 on day 6 showing immediate response of plankton biomass to fertilization (+P, P enrichment only; NP, N&P enrichment at 16:1 molar ratio; NNP, N&P enrichment at 75:1 molar ratio). (Photo credits: J. Elser)

Experiment 2: Experiment 2 sought to test the GRH at a larger scale and involved a whole-ecosystem manipulation with replicated internal controls coupled to metagenomic analyses. Briefly, five 40 cm transparent cylinders were inserted into the pond at the beginning of the experiment. Initial samples of the water column and surficial sediments were then taken in each of the five cylinders as well as at five locations in the pond itself. The pond was then fertilized with N and P at 16:1 ratio to achieve P concentration of 1 μM. SRP concentrations were then checked every 3–4 days, and the pond was re-enriched to return SRP to 1 μM. The experiment ran for 32 days. At 16 and 32 days, samples for analysis of seston C, N, and P concentrations were taken. At 32 days, samples of the water column were taken within each of the five enclosures and at five locations in the pond itself. (Sediment samples were also taken, but these were not analyzed.) DNA was extracted and then subjected to metagenomic sequencing of each replicate using an Illumina MiSeq.

Due to lack of read depth, two of the fertilized samples and one of the unfertilized samples were eliminated from further analyses. Thus, we had a total of seven metagenomes for analysis (four in the enriched pond, three in the unenriched enclosures). Sequences were analyzed for community composition, rRNA and tRNA gene copy numbers, genome size (using normalized core marker gene counts), and codon usage bias in ribosomal protein genes. These genomic traits constitute the "genomic signature" of growth (Vieira-Silva and Rocha 2010). Details of the analyses are presented in Okie et al. (in review).

Results

Experiment 1: Fertilization led to large and rapid changes in water column biomass. Indeed, changes were visible to the naked eye (Fig. 3.1d) within 6 days of the start of the experiment, but the magnitude of response decreased somewhat by day 42 following 21 days without enrichment (Fig. 3.2a and b). Adding N along with P produced a larger response in chlorophyll (days 6 and 21) and seston (day 21), but this synergistic effect had disappeared by day 42 (Fig. 3.2a and b). Nutrient enrichment also lowered seston C:P ratios in the +P and +NP16 treatments but not in the +NP75 treatment (data not shown; see Lee et al. 2015), suggesting that enrichment at an N:P ratio of 75 did not alleviate P limitation. Nutrient enrichment strongly reorganized the water column microbial structure (Fig. 3.2c); among the enrichment treatments, enrichment N:P ratio also had a secondary effect on community structure (Fig. 3.2c), promoting the growth of previously rare microbial taxa at the expense of the initially abundant, and potentially endemic, taxa (data not shown; see Lee et al. 2017). Importantly, qPCR data indicate that taxa with increased copy numbers of ribosomal RNA genes were stimulated in the +P treatment (Fig. 3.2d), consistent with the GRH. No changes in ribosomal RNA gene copy number were observed in the +NP16 and +NP75 treatments, suggesting the ecosystem P limitation was not sufficiently relieved for P-demanding, high copy number taxa to proliferate in those two treatments.

Experiment 2: Whole-pond fertilization with N and P led to large and rapid increases in chlorophyll and seston biomass in the water column (Fig. 3.3a) and increased seston P:C ratio (decreased C:P; Fig. 3.3c). Analyses of replicated metagenomic samples gave us an unusual opportunity to identify potential changes in microbial traits predicted by the GRH and its expansion, the "genomic signature" of growth. Fertilization did not change the relative proportions of major taxa in the community (i.e., *Bacteria*, Archaea, Eukarya, and viruses) although finer-scale changes in the abundance of various taxa led to major differences in the community composition of the fertilized pond relative to the unfertilized controls (Fig. 3.3b). Analysis of the replicated metagenomes resulted in consistent support for each component of the hypothesized genomic signature of growth (Fig. 3.3c). Members of the fertilized communities had consistently increased numbers of ribosomal RNA and tRNA genes, exhibited increased codon usage bias in the highly expressed

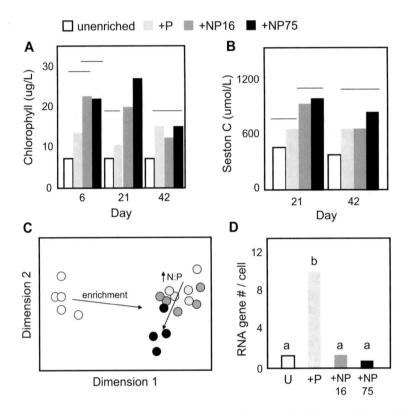

Fig. 3.2 Responses in Experiment 1. (**a**) Responses of autotrophic biomass (chlorophyll) to enrichment of P at different N:P ratios. Data are shown for sampling on days 6, 21, and 42 of the study. Horizontal bars join treatments that are not significantly different from each other. (**b**) Responses of total suspended organic C (seston) on days 21 and 42. (**c**) Multidimensional scaling results for the 16S community profiling data on day 21. Arrows indicate the overall effect of nutrient enrichment and the effects of increasing N:P ratio. (**d**) Cell number-normalized qPCR data for ribosomal RNA copy number on day 21. P enrichment along led to increased copy number, consistent with the GRH

ribosomal protein genes, and increased genome size. Larger genomes and increased numbers of rRNA and tRNA genes are consistent with increased production of P-rich ribosomes. Consequently, there is an observed increase in biomass P:C ratio in the fertilized treatment. It is not clear why N&P enrichment at 16:1 ratio was sufficient to shift ribosomal RNA copy numbers in this experiment but not in Experiment 1, when only the P alone treatment produced such a response.

Interpretations and Conclusions

The metagenomic analyses of our experiment provide some of the first field evidence supporting the predictions of an extended growth rate hypothesis in microbial assemblages. Fertilization consistently led to increased proliferation of taxa with a

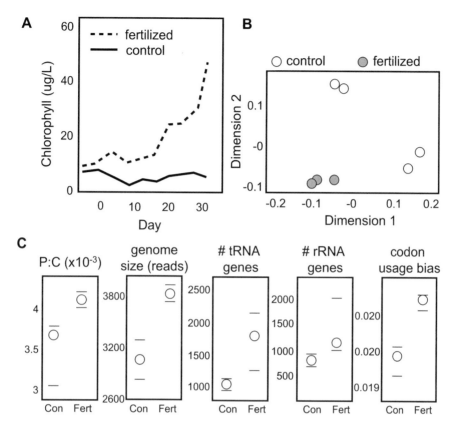

Fig. 3.3 Results of Experiment 2. (**a**) Response of autotroph biomass (chlorophyll) in Lagunita pond during fertilization (mean of five replicate samples) compared to control enclosures (means of five replicates). (**b**) Fertilization leads to major changes in microbial community structure (principal coordinates analysis) of 16S and 18S data from metagenomes. (**c**) Fertilization increased biomass P:C ratio and, consistent with the genomic signature of growth hypothesis, increased genome size, tRNA and rRNA gene copy number, and codon usage bias in the microbial community based on replicated metagenomic analysis. Circles indicate the treatment medians and bars indicate 75th and 25th percentile bounds. All responses were statistically significant ($p < 0.05$, one-tailed test)

P-intensive lifestyle driven by genomic features that support increased capacity for production and high-speed operation of protein synthesis machinery (increased dosages of ribosomal RNA and transfer RNA genes, increased codon usage bias). This appears to come at a cost in P competitiveness, as taxa favored by fertilization also had larger genomes (made of P-rich DNA) and higher biomass P:C ratios. Overall, we propose that these data are supportive of a view of microbial life histories as lying on a stoichiometric "r/K continuum" (Pianka 1972). On this continuum, "K-selected" species dominate under the normal, unenriched conditions in CCB, persisting with low-P biomass but modest capacities for growth. Upon nutrient enrichment, "r-selected" members of the rare biosphere with large capacity for growth and carrying P-rich protein synthesis machinery respond to sudden nutrient availability and become dominant, pushing CCB's characteristic microbial taxa to the background.

Our experiments confirm the strong nutrient sensitivity of the aquatic ecosystems of Cuatro Ciénegas, adding to data from similar nutrient enrichment studies involving stromatolites in the Rio Mesquites (Elser et al. 2005a, b and Chap. 6). Enrichment itself has a major impact in restructuring the communities (Fig. 3.2c) with a secondary impact of N:P stoichiometry. Importantly, this nutrient sensitivity places the unique and potentially endemic microbial taxa at particular risk, as these are the taxa that have an extended evolutionary history under the low-nutrient, stoichiometrically imbalanced nutrient conditions in the basin and that seem to be ill-equipped for life under high nutrient conditions. This argues for strict management approaches that prevent inputs of N and P from human and livestock waste and from transport (aerial or waterborne) of agricultural fertilizers. Of particular concern is long-distance nutrient transport from industrial/urban sources (e.g., NO_x compounds) and from agricultural fertilizer and fertilized soils in wind-borne dust, as these cannot be managed at the local scale but must be addressed regionally.

References

Condon C, Liveris D, Squires C et al (1995) Ribosomal RNA operon multiplicity in Escherichia coli and the physiological implications of rrn inactivation. J Bacteriol 177:4152–4156

Downing JA, McCauley E (1992) The nitrogen: phosphorus relationship in lakes. Limnol Oceanogr 37:936–945. https://doi.org/10.4319/lo.1992.37.5.0936

Elliott TA, Gregory TR (2015) What's in a genome? The C-value enigma and the evolution of eukaryotic genome content. Philos Trans R Soc Lond B Biol Sci 370(1678):20140331. https://doi.org/10.1098/rstb.2014.0331

Elser JJ, Dobberfuhl DR, MacKay NA et al (1996) Organism size, life history, and N:P stoichiometry. Bioscience 46:674–684. https://doi.org/10.2307/1312897

Elser JJ, O'Brien J, Dobberfuhl DR (2000a) The evolution of ecosystem processes: growth rate and elemental stoichiometry of a key herbivore in temperate and arctic habitats. J Evol Biol 13:845–853. https://doi.org/10.1046/j.1420-9101.2000.00215.x

Elser JJ, Sterner RW, Gorokhova E et al (2000b) Biological stoichiometry from genes to ecosystems. Ecol Lett 3:540–550. https://doi.org/10.1111/j.1461-0248.2000.00185.x

Elser JJ, Schampel JH, Garcia-Pichel F et al (2005a) Effects of phosphorus enrichment and grazing snails on modern stromatolitic microbial communities. Freshw Biol 50:1808–1825. https://doi.org/10.1111/J.1365-2427.2005.01451.X

Elser JJ, Schampel JH, Kyle M et al (2005b) Response of grazing snails to phosphorus enrichment of modern stromatolitic microbial communities. Freshw Biol 50:1826–1835. https://doi.org/10.1111/j.1365-2427.2005.01453.x

Elser JJ, Bracken MES, Cleland EE et al (2007) Global analysis of nitrogen and phosphorus limitation of primary producers in freshwater, marine and terrestrial ecosystems. Ecol Lett 10:1135–1142. https://doi.org/10.1111/J.1461-0248.2007.01113.X

Elser JJ, Acquisti C, Kumar S (2011) Stoichiogenomics: the evolutionary ecology of macromolecular elemental composition. Trends Ecol Evol 26:38–44 https://doi.org/10.1016/j.tree.2010.10.006

Fagan WF, Siemann E, Mitter C et al (2002) Nitrogen in insects: implications for trophic complexity and species diversification. Am Nat 160:784–802. https://doi.org/10.1086/343879

Fu R, Gong J (2017) Single cell analysis linking ribosomal (r)DNA and rRNA copy numbers to cell size and growth rate provides insights into molecular protistan ecology. J Eukaryot Microbiol 64:885–896. https://doi.org/10.1111/jeu.12425

Gyorfy Z, Draskovits G, Vernyik V et al (2015) Engineered ribosomal RNA operon copy-number variants of E. coli reveal the evolutionary trade-offs shaping rRNA operon number. Nucleic Acids Res 43:1783–1794. https://doi.org/10.1093/nar/gkv040

Hartman WH, Ye R, Horwath WR et al (2017) A genomic perspective on stoichiometric regulation of soil carbon cycling. ISME J 11:2652–2665. https://doi.org/10.1038/ismej.2017.115

Hessen DO, Elser JJ, Sterner RW et al (2013) Ecological stoichiometry: an elementary approach using basic principles. Limnol Oceanogr 58:2219–2236. https://doi.org/10.4319/lo.2013.58.6.2219

Klappenbach JA, Dunbar JM, Schmidt TM (2000) rRNA operon copy number reflects ecological strategies of bacteria. Appl Environ Microbiol 66:1328–1333. https://doi.org/10.1128/aem.66.4.1328-1333.2000

Lauro FM, McDougald D, Thomas T et al (2009) The genomic basis of trophic strategy in marine bacteria. PNAS 106:15527–15533. https://doi.org/10.1073/pnas.0903507106

Lee ZM-P, Bussema C, Schmidt TM (2009) rrnDB: documenting the number of rRNA and tRNA genes in bacteria and archaea. Nucleic Acids Res 37(Suppl 1):D489–D493. https://doi.org/10.1093/nar/gkn689

Lee ZM, Steger L, Corman JR et al (2015) Response of a stoichiometrically imbalanced ecosystem to manipulation of nutrient supplies and ratios. PLoS One 10(4):e0123949

Lee ZM-P, Poret-Peterson AT, Siefert JL et al (2017) Nutrient stoichiometry shapes microbial community structure in an evaporitic shallow pond. Front Microbiol 8:949. https://doi.org/10.3389/fmicb.2017.00949

Maeda M, Shimada T, Ishihama A (2016) Strength and regulation of seven rRNA promoters in Escherichia coli. PLoS One 10(12):e0144697. https://doi.org/10.1371/journal.pone.0144697

Nemergut DR, Knelman JE, Ferrenberg S et al (2016) Decreases in average bacterial community rRNA operon copy number during succession. ISME J 10:1147–1156. https://doi.org/10.1038/ismej.2015.191

Paul BJ, Ross W, Gaal T, Gourse RL (2004) rRNA transcription in Escherichia coli. Annu Rev Genet 38:749–770. https://doi.org/10.1146/annurev.genet.38.072902.091347

Pianka ER (1972) r and K selection or b and d selection? Am Nat 106:581–588. https://doi.org/10.1086/282798

Prokopowich CD, Gregory TR, Crease TJ (2003) The correlation between rDNA copy number and genome size in eukaryotes. Genome 46:48–50. https://doi.org/10.1139/g02-103

Rocha EPC (2004) Codon usage bias from tRNA's point of view: redundancy, specialization, and efficient decoding for translation optimization. Genome Res 14:2279–2286. https://doi.org/10.1101/gr.2896904

Roller BRK, Stoddard SF, Schmidt TM (2016) Exploiting rRNA operon copy number to investigate bacterial reproductive strategies. Nat Microbiol 1:16160. https://doi.org/10.1038/nmicrobiol.2016.160

Vieira-Silva S, Rocha EPC (2010) The systemic imprint of growth and its uses in ecological (meta) genomics. PLoS Genet 6(1):e1000808. https://doi.org/10.1371/journal.pgen.1000808

Wagner R (1994) The regulation of ribosomal RNA synthesis and bacterial cell growth. Arch Microbiol 161:100–109. https://doi.org/10.1007/bf00276469

Woods HA, Fagan WF, Elser JJ et al (2004) Allometric and phylogenetic variation in insect phosphorus content. Funct Ecol 18:103–109. https://doi.org/10.1111/j.1365-2435.2004.00823.x

Yano K, Wada T, Suzuki S et al (2013) Multiple rRNA operons are essential for efficient cell growth and sporulation as well as outgrowth in Bacillus subtilis. Microbiology 159:2225–2236. https://doi.org/10.1099/mic.0.067025-0

Chapter 4
The Effect of Nutrient Availability on the Ecological Role of Filamentous Microfungi: Lessons from Elemental Stoichiometry

Yunuen Tapia-Torres, Patricia Vélez, Felipe García-Oliva, Luis E. Eguiarte, and Valeria Souza

Contents

Abstract Ecological stoichiometry theory helps us to better understand trophic interactions by analyzing the imbalances in the relative supplies of key elements (carbon, nitrogen, and phosphorus) between organisms and their resources. However, the mechanisms that control elemental stoichiometry in different taxonomic groups and the effects of nutrient supply imbalances are not yet clear. Aquatic microfungi are an ecological group of microorganisms ranging from those adapted

Y. Tapia-Torres (✉)
Escuela Nacional de Estudios Superiores Unidad Morelia, Universidad Nacional Autónoma de México, Morelia, Mexico
e-mail: ytapia@enesmorelia.unam.mx

P. Vélez
Departamento de Ecología Evolutiva, Instituto de Ecología, Universidad Nacional Autónoma de México, Coyoacan, Mexico

Departamento de Botánica, Instituto de Biología, Universidad Nacional Autónoma de México, Coyoacan, Mexico

F. García-Oliva
Instituto de Investigaciones en Ecosistemas y Sustentabilidad, Universidad Nacional Autónoma de México, Morelia, Mexico

L. E. Eguiarte · V. Souza
Departamento de Ecología Evolutiva, Instituto de Ecología, Universidad Nacional Autónoma de México, Coyoacan, Mexico

© Springer International Publishing AG, part of Springer Nature 2018
F. García-Oliva et al. (eds.), *Ecosystem Ecology and Geochemistry of Cuatro Cienegas*, Cuatro Ciénegas Basin: An Endangered Hyperdiverse Oasis, https://doi.org/10.1007/978-3-319-95855-2_4

to complete their life cycles in aquatic habitats to those that occurring in water fortuitously. Aquatic fungi are important regulators of plant productivity, community dynamics, and diversity in nutrient-poor and extreme ecosystems. Because aquatic fungi are heterotrophs, it has been assumed that they possess high degree of stoichiometric homeostasis. However, data concerning their elemental composition and their degree of homeostasis remain scarce. Herein, we analyzed the C:N:P stoichiometry of mycelia in ten aquatic microfungi isolated from three hydrological systems with different nutrient conditions within Cuatro Cienegas Basin (CCB). Our hypotheses were (a) variations in C:N:P ratios reflect divergent life history strategies between the three environments, independently of the fungal taxa involved, and (b) C:N:P ratios reflect physiological adjustments associated with specific taxa, independent of the environmental characteristics. Our results provide some support for the first hypothesis, as the apparent capacity for elemental stoichiometry regulation in the aquatic microfungi was not linked to phylogenetic relationships but appeared to be an adaptation to resource availabilities in the environment in which they grew. The microfungi isolated from the most oligotrophic site within the CCB (Pozas Rojas) most strongly regulated their elemental stoichiometry in comparison with fungal isolates from other sites within CCB.

Keywords Aquatic microfungi · C:N:P stoichiometry · Ecological stoichiometry · Fungi phylogenetic analyses

Introduction

Microbial communities regulate core ecosystem processes (e.g., organic matter decomposition, soil carbon (C) sequestration, and nutrient (nitrogen (N), phosphorus (P), etc.) transformations) and thus are an essential component of global biogeochemical cycles (Schimel and Bennett 2004). Fungi, as a large and diverse component of microbial diversity in soils (Fierer et al. 2007), fulfill a major role in nutrient fluxes, soil development, and decomposition rates and thus strongly affect the uptake of nutrients by plants and other organisms (Dighton 1997; Schneider et al. 2005). However, despite their importance in nutrient cycling, underlying information on fungal-based processes associated with nutrient dynamics in the majority of ecosystems remains incomplete.

Use of ecological stoichiometry theory can improve our understanding on trophic dynamics relationships within a system by analyzing the imbalances in the supply of elements (C:N:P) between organisms and their resources (Sterner and Elser 2002). Different studies have shown that energy and nutrient imbalances (C:N, C:P, or N:P) between consumers and their resources strongly constrain ecosystem nutrient cycling and limit consumer reproduction and growth (Andersen et al. 2004; Frost et al. 2005; Persson et al. 2010). Key to nutrient recycling and food qualify effects of stoichiometry is the observation that different organisms vary in how they regulate their elemental composition in response to variation in resource stoichiometry; this regulation is called "stoichiometric homeostasis" (Sterner and

Elser 2002). This concept is central in the development of the ecological stoichiometry framework (Elser and Urabe 1999; Koojiman 1995; Sterner and Elser 2002) and therefore also to biological stoichiometry (Elser et al. 2000a). Generally, heterotrophs are thought to have higher degree of stoichiometric homeostasis than autotrophs (Sterner and Elser 2002; Persson et al. 2010). This difference may reflect the nutritional environments experienced by the different categories of organisms. For example, variations in the stoichiometric composition of autotrophic organisms may be triggered by wide fluctuations in light, nutrient supplies, and growth rates (Sterner et al. 1998); meanwhile, heterotrophs experience a far more muted range of resource stoichiometry in their diets (Andersen and Hessen 1991). Recently, however, this generality has been questioned by several authors and has resulted in an effort to understand in more detail the mechanisms that control elemental stoichiometry of biomass and its regulation in different taxonomic groups (Danger and Chauvet 2013; Persson et al. 2010; Zhang and Elser 2017).

Despite the fundamental role of fungi in organic matter decomposition and nutrient cycling in both aquatic and terrestrial ecosystems, few data are currently available on fungal elemental composition and regulation. Recently, Zhang and Elser (2017) conducted a meta-analysis of C:N:P fungal stoichiometry and found that the fungal C:N:P stoichiometry is, on average, 250:16:1 (by atoms). Note that the canonical "Redfield ratio" for marine planktonic biomass is 106:16:1(Redfield 1958), while Cleveland and Liptzin (2007) reported a value of 60:7:1 for soil bacteria. It is notable to observe that the fungal N:P ratio proposed by Zhang and Elser (2017) is very similar to the one proposed by Redfield (1958). However, Cleveland and Lipzin's study suggests that bacteria seem to be more P-rich, on average. The variation in elemental composition is linked to the differences in organismal allocation to biochemical and structural components that differ in C, N, and P contents (Sterner and Elser 2002). For example, it has been reported that in microorganisms, from 50% to 90% of the total P in biomass is associated with P in rRNA, while in larger organisms this percentage can be reduced to <10%, especially in vertebrates that contain most of their P in bone (Sterner and Elser 2002). While a variety of data have accumulated to better illuminate these patterns and their underlying drivers, we lack studies that examine the degree to which fungal C:N:P stoichiometry in natural communities is driven by physiological adjustment of existing taxa to variation in external supplies or by shifts in the relative abundance of taxa that differ in taxon-specific C:N:P stoichiometry.

Oligotrophic environments are natural laboratories that test these possibilities as they make it possible to study taxonomically different organisms with metabolic capabilities adapted to the low nutrient and energy availability. CCB is an oligotrophic isolated basin in the Chihuahuan Desert, Mexico, encompassing heterogeneous aquatic systems (Minckley and Cole 1968), which, despite their low phosphorus content (Breitbart et al. 2009; Elser et al. 2005; Peimbert et al. 2012), harbor varied and diverse microbial communities (Escalante et al. 2008; Souza et al. 2006; Velez et al. 2016). Prokaryotic communities in the CCB have been reported to follow global ecoenzymatic stoichiometry patterns (1:1:1), being differentially co-limited by C and either by N or P depending on where they occur in the basin (Lee et al.

2015; Tapia-Torres et al. 2015). It has also been observed that bacterial communities are able to obtain P from substrates with different oxidation states (Tapia-Torres et al. 2016). However, no data are currently available on fungal elemental composition in CCB, despite their central role in detritus decomposition and biogeochemical cycling.

Aquatic fungi are a diverse group of microorganisms ranging from those adapted to complete their life cycles in aquatic habitats to those that occur in water fortuitously by being washed or blown (transient or facultative; Shearer et al. 2007). Although these heterotrophs have been recognized as important regulators of plant productivity, community dynamics, and diversity in nutrient-poor and extreme ecosystems, data concerning their elemental composition and their degree of homeostasis remain scarce (Persson et al. 2010; Van Der Heijden et al. 2008). Some reports suggest that fungal mycelia possess higher C:nutrient ratios in relation to bacteria as they are able to assimilate nutrients from the water column to overcome the high C:nutrient ratios that are typical of dead organic matter (Cheever et al. 2012; Cross et al. 2007; Pastor et al. 2014; Scott et al. 2013). Remarkably, fungal elemental molar ratios have been reported to range from 21 to 2800 (C:P), 3.3 to 220 (C:N), and 1.5 to 140 (N:P) (Danger and Chauvet 2013; Elser et al. 2000b; Persson et al. 2010; Zhang and Elser 2017).

While Zhang and Elser (2017) showed that fungal N:P ratios can vary with geographic position (latitude) and with environmental conditions (particularly precipitation and temperature), most of the data included in their study involve reproductive structures of *Ascomycota* and *Basidiomycota*, overlooking, due to the lack of published data, the stoichiometry of aquatic fungal mycelium under in situ growth conditions. Thus, more work is needed to better understand fungal stoichiometry and its regulatory mechanisms. To continue advancing in the understanding of the elemental stoichiometry and the ecological role of fungi, our main objective was to analyze the C:N:P stoichiometry in the mycelium of ten aquatic microfungi isolated from three hydrological systems with naturally different elemental stoichiometry within CCB. Our interest was focused on evaluating the role of environmental conditions versus phylogenetic influences in influencing C:N:P ratios of microfungal mycelia in these contrasting freshwater systems. So, we tested if variations in C:N:P ratios reflect divergent life history strategies between the three environments, independent of the fungal taxa, or if C:N:P ratios reflected physiological adjustments associated with specific taxa, regardless of environmental characteristics.

Materials and Methods

This study was conducted using ten fungal isolates obtained by Velez et al. (2016) from three hydrological systems within the Cuatro Cienegas Basin (CCB; 26° 50′ N and 102° 8′ W) in the central region of the Chihuahuan Desert, Coahuila, México: Churince, Becerra, and Pozas Rojas. Churince (26° 50′ N; 102° 8′ W) and Becerra (26° 52′ N; 102° 8′ W) are part of a hydrological system that is located in the west region of the basin and consists of a perennial spring and two desiccation lagoons that are connected by short shallow streams (Lopez-Lozano et al. 2012) that are rich in sulfates (Minckley and Cole 1968; Tobler and Carson 2010). On the other hand, the Pozas Rojas system is located at the east of the valley (26° 50' N; 102° 1' W) and involves several small desiccation ponds rich in carbonates with conditions that fluctuate drastically both within and across years (Minckley and Cole 1968; Tobler and Carson 2010).

Chemical Analyses of Mycelia, Water Samples, and Potato Dextrose Broth

Mycelia of each taxon were grown in triplicate at 25 °C in the dark in potato dextrose broth (PDA; Fluka Analytical Lot # SLBJ7991V) for 15 days. Mycelium biomass was harvested and rinsed twice with sterile deionized water before being recovered onto a pre-weighed GF/F filter (Whatman International Ltd., Maidstone, UK). Filters were then dried at 40 °C for 4 days. The elemental composition of mycelium biomass was measured on agate mortar ground subsamples.

We collected water samples from each of the three hydrological systems from which fungi were isolated. For each, concentrations of total organic carbon (TOC), total nitrogen (TN), and total phosphorus (TP) were determined. Additionally, we quantified TOC, TN, and TP in the PDA medium in which mycelia of each taxon were grown for each mycelium biomass sample.

Before the analysis of total nutrient forms, water samples were filtered through a 0.45 μm Millipore filter. Total C and inorganic C were determined by combustion and coulometric detection (Huffman 1977) respectively using a total carbon analyzer (UIC Mod. CM5012; Chicago, IL, USA). Total organic C was calculated as the difference between total C and inorganic C. Total N and total P were determined after acid digestion. Total N was determined by a macro-Kjeldahl method with colorimetric determination (Bremner and Mulvaney 1982), and total P was determined by a molybdate-based colorimetric method after reduction with ascorbic acid (Murphy and Riley 1962). N and P forms were determined using a Bran-Luebbe Auto Analyzer III (Norderstedt, Germany).

The taxonomic assignation of each isolate is described in Velez et al. (2016).

Results and Discussion

The overall elemental stoichiometry of the water of the three hydrological systems (Pozas Rojas, Becerra, and Churince) differed strongly: 15820:158:1, 1213:6:1, and 663:4:1 (mass), respectively. These stoichiometric ratios reflect the relative resource availabilities in natural conditions experienced by the aquatic microfungal isolates analyzed. The Pozas Rojas system had lower P contents and higher stoichiometric

Table 4.1 Total nutrient (mg/l) of the three hydrological systems (Pozas Rojas, Becerra, and Churince) in CCB, Mexico, and of the potato dextrose broth where mycelia of each taxon were grown

Site	TOC[a]	TN[b]	TP[c]	TOC:TN	TOC:TP	TN:TP	C:N:P
Pozas Rojas	9492	94.3	0.6	101	15,820	158	15,820:158:1
Becerra	58.21	0.28	0.048	208	1213	6	1213:6:1
Churince	65.6	0.375	0.099	175	663	4	663:4:1
Potato dextrose broth	8380	129	16.56	65	506	8	506:8:1

[a]Total organic carbon
[b]Total nitrogen
[c]Total phosphorus

imbalance in relation to C and N than the other two hydrological systems (Table 4.1), confirming results reported previously (Peimbert et al. 2012). Additionally, the media in which the fungi were grown in the laboratory (PDA medium) had elemental stoichiometry (506:8:1) that was quite distinct from the three hydrological systems (Table 4.1). After 15 days of growth in PDA in the laboratory, we observed an average fungal C:N:P stoichiometry of 38:3:1 (molar) that differs considerably from the 250:16:1 average reported by Zhang and Elser (2017). A wide range of elemental ratios has been reported in fungi (Elser et al. 2000b; Persson et al. 2010; Danger and Chauvet 2013; Zhang and Elser 2017), varying from 21 to 2800 (C:P), 3.3 to 220 (C:N), and 1.5 to 140 (N:P). Our data are within the lower part of this range. These data suggest that, even though the assessed aquatic microfungi are found in oligotrophic systems with very high C:P, C:N, and N:P ratios, they possess considerable capacity to assimilate nutrients and produce an elemental stoichiometry similar to that of organisms that live in more nutrient-rich environments.

These results also suggest that, in contrast to the local endemic *Bacillus* with its very strong stoichiometric bias and low nutrient contents (Valdivia-Anistro et al. 2016), fungi may be just "passing by" in the Cuatro Cienegas systems and retain a high-nutrient lifestyle. Indeed, this group includes many members that have been typically regarded as terrestrial fungi that occasionally occur in freshwater systems (Duarte et al. 2015), suggesting a broader ecological niche. Furthermore, it is important to note that the CCB's freshwater springs are open systems, where the input of external material such as soil or plant remains (rich in lignin and cellulose, translating to C and N) is likely. These findings also highlight the possibility of material fluxes across the terrestrial and aquatic interface of the set of desiccation small springs in Pozas Rojas that shrink in summer and expand as evaporation decreases in winter.

Theory predicts that transient taxa might undergo selection processes after a considerable time lag since they cannot predict the when or the where of the "next jump" to a new environment. It is possible that these wide-ranging microbes have been selected to be capable of successfully proliferating under rapidly fluctuating conditions as they transit from terrestrial to aquatic systems (Velez et al. 2018). We suggest that the aquatic microfungal isolates that we analyzed were using nutrients, and especially P, inefficiently (as reflected in low N:P and especially C:P

Table 4.2 Analyzed microfungal isolates

Isolate	Phylum	OTU	Accession number	Isolation site
4	*Ascomycota*	*Cladosporium* sp. 1	KU933775	Pozas Rojas
7	*Zygomycota*	*Rhizopus arrhizus*	KU933760	Pozas Rojas
18	*Ascomycota*	*Plectosphaerella* sp.	KU933707	Pozas Rojas
AS	*Ascomycota*	*Cladosporium* sp. 1	KU933777	Pozas Rojas
CK	*Ascomycota*	*Cladosporium* sp. 2	KU933725	Pozas Rojas
CÑ	*Basidiomycota*	Agaricales sp. 2	KU933730	Pozas Rojas
DS	*Ascomycota*	*Myrothecium* sp. 2	KU933731	Becerra
AJ	*Ascomycota*	*Pyrenochaetopsis* sp.	KU933764	Churince
EG	*Ascomycota*	*Myrothecium* sp. 2	KU933738	Churince
MM	*Ascomycota*	Pleosporales sp. 1	KU933815	Churince

Accession numbers correspond to Velez et al. (2016)

ratios) and did not show local adaptation to the low P availability and imbalanced N:P stoichiometry in CCB. However, further analyses involving more careful culture conditions across a range of media conditions and growth rates are needed that allow us to corroborate this conclusion.

Our isolates belong to three different phyla: *Ascomycota* (80%), *Basidiomycota* (10%), and *Zygomycota* (10%) (Table 4.2; Fig. 4.1). However, despite the fact that our fungal isolates are distributed in the three mayor phyla, the differences observed in fungal elemental stoichiometry were not associated with the taxonomic group (Table 4.3). Instead, they were generally associated with the stoichiometric imbalance of the isolation sites (Fig. 4.2). We observed lower fungal C:N, C:P, and N:P ratios in the isolates from Churince (16:2:1 molar) than in the isolates from Pozas Rojas (40:3:1 molar). These suggest weak P limitation in fungi from Pozas Rojas, while N may represent the principal limitation in Churince. These results agree with evidence from soil microbial communities at CCB in which the prokaryotic community in the Churince system was co-limited by C and N, while, in the Pozas Azules (adjacent to the Pozas Rojas), microbial communities were co-limited by C and P (Tapia-Torres et al. 2015).

Conclusions

Our data suggest that the aquatic microfungal isolates that we analyzed showed some signs of local adaptation that allows them to live in an imbalanced stoichiometry present in the water pools (Fig. 4.2). The C:N:P ratios of mycelia grown in PDA were lower compared to the ratios reported by Zhang and Elser (2017), which suggests that they are highly plastic in stoichiometry as we would expect that the fungi have higher C:P and N:P ratios under in situ conditions. However, more work involving more media C:N:P ratios is needed. Also more studies about their C:N:P

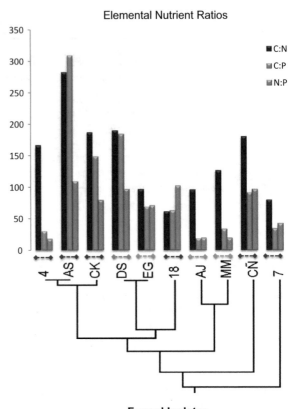

Fig. 4.1 Elemental nutrient ratios (C:N, C:P, and N:P) of each isolate and the hierarchical clustering (UPGMA) showing genetic distances among the tested isolates. The dotted lines in different color represent the site of isolation: Pozas Rojas (red), Churince (green), and Becerra (blue). C:N and C:P values were multiplied by 10 to plot

Table 4.3 Analyzed fungal isolates: carbon (C), nitrogen (N), and phosphorus (P) values given in mg/g

Isolate	OTU	C	N	P	C:N	C:P	N:P	C:N:P
4	*Cladosporium* sp. 1	361	21.7	11.9	16.6	30.2	1.82	30:1.82:1
7	*Rhizopus arrhizus*	339	42.2	9.9	8	34.3	4.27	34:4.27:1
18	*Plectosphaerella* sp.	460	75.0	7.9	6.1	62.8	10.2	63:10.2:1
AS	*Cladosporium* sp. 1	486	17.2	1.6	28.2	309	10.9	309:10.9:1
CK	*Cladosporium* sp. 2	462	24.7	3.1	18.7	149	8	149:8:1
CÑ	*Agaricales* sp. 2	389	21.6	4.3	18	91.6	9.7	184:9.7:1
DC	*Myrothecium* sp. 2	448	23.6	2.4	19	184	9.7	184:9.7:1
AJ	*Pyrenochaetopsis* sp.	402	41.9	21.5	9.6	18.8	2	19:2:1
EG	*Myrothecium* sp. 2	435	45.0	6.3	9.7	68.7	7	69:7:1
MM	*Pleosporales* sp. 1	280	22.2	8.4	12.6	33.5	2.7	34:2.7:1

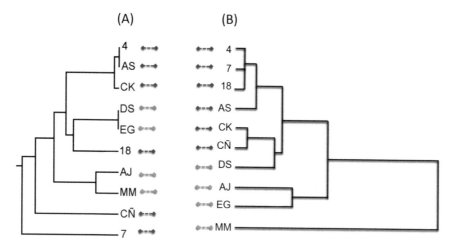

Fig. 4.2 Hierarchical clustering (UPGMA) showing genetic distances among the tested isolates (**a**) and Euclidean distances associated to elemental stoichiometry C:N, C:P, and N:P (**b**). The dotted lines in different color represent the site of isolation: Pozas Rojas (red), Churince (green), and Becerra (blue)

stoichiometry in situ conditions are needed to confirm that our observations represent a plastic response rather than selection of rare low C:P and N:P strains from the mixed community in the pools.

Additionally, despite the apparent plasticity in response to the growth medium, habitat of origin (Pozas Rojas vs Churince) was a strong predictor of their C:N:P ratios, suggesting an environmental filtering of species by habitat C:N:P (Fig. 4.2). However, detailed work of fungal diversity in these systems is needed to better establish our findings.

References

Andersen T, Hessen DO (1991) Carbon, nitrogen, and phosphorus content of freshwater zooplankton. Limnol Oceanogr 36:807–814

Andersen T, Elser JJ, Hessen DO (2004) Stoichiometry and population dynamics. Ecol Lett 7:884–900

Breitbart M, Hoare A, Nitti A (2009) Metagenomic and stable isotopic analyses of modern freshwater microbialites in Cuatro Cienegas, Mexico. Environ Microbiol 11:16–34. https://doi.org/10.1111/j.1462-2920.2008.01725.x

Bremner JM, Mulvaney CS (1982) Nitrogen-total. In: Page AL, Miller RH, Keeney DR (eds) Methods of soil analysis: chemical and microbiological properties. American Society of Agronomy and Soil Science Society of America, Madison, WI, pp 595–624

Cheever BM, Kratzer EB, Webster JR (2012) Immobilization and mineralization of N and P by heterotrophic microbes during leaf decomposition. Freshw Sci 31:133–147

Cleveland CC, Liptzin D (2007) C:N:P stoichiometry in soil: is there a "Redfield ratio" for the microbial biomass? Biogeochemistry 85:235–252

Cross WF, Wallace JB, Rosemond AD (2007) Nutrient enrichment reduces constraints on material flows in a detritus-based food web. Ecology 88:2563–2575. https://doi.org/10.1890/06-1348.1

Danger M, Chauvet E (2013) Elemental composition and degree of homeostasis of fungi: are aquatic hyphomycetes more like metazoans, bacteria or plants? Fungal Ecol 6:453–457

Dighton J (1997) Nutrient cycling by saprotrophic fungi in terrestrial habitats. Mycota 4:271–279

Duarte S, Bärlocher F, Trabulo J et al (2015) Stream-dwelling fungal decomposer communities along a gradient of eutrophication unraveled by 454 pyrosequencing. Fungal Divers 70:127–148

Elser JJ, Urabe J (1999) The stoichiometry of consumer-driven nutrient recycling: theory, observations and consequences. Ecology 80:735–751

Elser JJ, Sterner RW, Gorokhova E et al (2000a) Biological stoichiometry from genes to ecosystems. Ecol Lett 3:540–550. https://doi.org/10.1111/j.1461-0248.2000.00185.x

Elser JJ, O'Brien WJ, Dobberfuhl DR et al (2000b) The evolution of ecosystem processes: growth rate and elemental stoichiometry of a key herbivore in temperate and arctic habitats. J Evol Biol 13:845–853. https://doi.org/10.1046/j.1420-9101.2000.00215.x

Elser JJ, Schampel JH, Garcia-Pichel F et al (2005) Effects of phosphorus enrichment and grazing snails on modern stromatolitic microbial communities. Freshw Biol 50:1808–1825. https://doi.org/10.1111/j.1365-2427.2005.01451.x

Escalante AE, Eguiarte LE, Espinosa-Asuar L et al (2008) Diversity of aquatic prokaryotic communities in the Cuatro Cienegas basin. FEMS Microbiol Ecol 65:50–60. https://doi.org/10.1111/j.1574-6941.2008.00496.x

Fierer N, Bradford MA, Jackson RB (2007) Toward an ecological classification of soil bacteria. Ecology 88:1354–1364. https://doi.org/10.1890/05-1839

Frost PC, Michelle A, Evans-White Z et al (2005) Are you what you eat? Physiological constraints on organismal stoichiometry in an elementally imbalanced world. Oikos 109:18–28

Huffman EWD (1977) Performance of a new automatic carbon dioxide coulometer. Microcheml J 22(4):567–573

Koojiman SALM (1995) The stoichiometry of animal energetics. J Theor Biol 177:139–149

Lee ZM, Steger L, Corman JR et al (2015) Response of a stoichiometrically imbalanced ecosystem to manipulation of nutrient supplies and ratios. PLoS One 10(4):e0123949. https://doi.org/10.1371/journal.pone.0123949

López-Lozano NE, Eguiarte LE, Bonilla-Rosso G et al (2012) Bacterial communities and the nitrogen cycle in the gypsum soil of Cuatro Ciénegas Basin, Coahuila: a Mars analogue. Astrobiology 12:699–709. https://doi.org/10.1089/ast.2012.0840

Minckley WL, Cole GA (1968) Preliminary limnologic information on waters of the Cuatro Cienegas Basin, Mexico. Southwest Nat 13:421–431. https://doi.org/10.2307/3668909

Murphy J, Riley JP (1962) A modified single solution method for the determination of phosphorus in natural water. Anal Chim Acta 27:31–36

Pastor A, Compson ZG, Dijkstra P et al (2014) Stream carbon and nitrogen supplements during leaf litter decomposition: contrasting patterns for two foundation species. Oecologia 176:1111–1121. https://doi.org/10.1007/s00442-014-3063-y

Peimbert M, Alcaraz LD, Bonilla-Rosso G et al (2012) Comparative metagenomics of two microbial mats at Cuatro Ciénegas Basin I: ancient lessons on how to cope with an environment under severe nutrient stress. Astrobiology 12:648–658. https://doi.org/10.1089/ast.2011.0694

Persson J, Fink P, Goto A et al (2010) To be or not to be what you eat: regulation of stoichiometric homeostasis among autotrophs and heterotrophs. Oikos 119:741–751. https://doi.org/10.1111/j.1600-0706.2009.18545.x

Redfield AC (1958) The biological control of chemical factors in the environment. Am Sci 46:205–221

Schimel JP, Bennett J (2004) Nitrogen mineralization: challenges of a changing paradigm. Ecology 85:591–602. https://doi.org/10.1890/03-8002

Schneider K, Renker C, Maraun M (2005) Oribatid mite (Acari, Oribatida) feeding on ectomycorrhizal fungi. Mycorrhiza 16:67–72

Scott EE, Prater C, Norman E et al (2013) Leaf-litter stoichiometry is affected by streamwater phosphorus concentrations and litter type. Freshw Sci 32:753–761. https://doi.org/10.1899/12-215.1

Shearer CA, Descals E, Kohlmeyer B et al (2007) Fungal biodiversity in aquatic habitats. Biodivers Conserv 16:49–67. https://doi.org/10.1007/s10531-006-9120-z

Souza V, Espinosa-Asuar L, Escalante AE et al (2006) An endangered oasis of aquatic microbial biodiversity in the Chihuahuan desert. PNAS 103:6565–6570. https://doi.org/10.1073/pnas.0601434103

Sterner RW, Elser JJ (2002) Ecological stoichiometry: the biology of elements from molecules to the biosphere. Princeton University Press, Princeton, NJ

Sterner RW, Clasen L et al (1998) Carbon: phosphorus stoichiometry and food chain production. Ecol Lett 1:146–150

Tapia-Torres Y, Elser JJ, Souza V et al (2015) Ecoenzymatic stoichiometry at the extremes: how microbes cope in an ultra-oligotrophic desert soil. Soil Biol Biochem 87:34–42 https://doi.org/10.1016/j.soilbio.2015.04.007

Tapia-Torres Y, Rodríguez-Torres D, Islas A et al (2016) How to live with phosphorus scarcity in soil and sediments: lessons from bacteria. Appl Environ Microbiol 82:4652–4662

Tobler M, Carson EW (2010) Environmental variation, hybridization, and phenotypic diversification in Cuatro Cienegas pupfishes. J Evol Biol 23:1475–1489

Valdivia-Anistro JA, Eguiarte-Fruns LE, Delgado-Sapién G et al (2016) Variability of rRNA operon copy number and growth rate dynamics of Bacillus isolated from an extremely oligotrophic aquatic ecosystem. Front Microbiol 6:1486. https://doi.org/10.3389/fmicb.2015.01486

Van Der Heijden MG, Bardgett RD, Van Straalen NM (2008) The unseen majority: soil microbes as drivers of plant diversity and productivity in terrestrial ecosystems. Ecol Lett 11:296–310. https://doi.org/10.1111/j.1461-0248.2007.01139.x

Velez P, Gasca-Pineda J, Rosique-Gil E et al (2016) Microfungal oasis in an oligotrophic desert: diversity patterns and community structure in three freshwater systems of Cuatro Ciénegas, Mexico. PeerJ 4:e2064

Velez P, Espinosa-Asuar L, Figueroa M et al (2018 Under revision) Nutrient dependent cross-kingdom interactions: microfungi and bacteria from an oligotrophic desert oasis. Front Microbiol (Under revision)

Zhang J, Elser JJ (2017) Carbon:nitrogen:phosphorus stoichiometry in fungi: a meta-analysis. Front Microbiol 8:1281. https://doi.org/10.3389/fmicb.2017.01281

Chapter 5
Life on a Stoichiometric Knife Edge: Biogeochemical Interactions and Trophic Interactions in Stromatolites in Rio Mesquites

Jessica Corman and James Elser

Contents

Abstract One of the unique features of CCB is its diverse and abundant microbialites ("stromatolites"). For example, in Rio Mesquites there are abundant spheroid or "oncoid" microbialites (akin to fossil stromatolites). These microbial communities form layers of $CaCO_3$ when photosynthesis raises pH and precipitates $CaCO_3$. We hypothesized that $CaCO_3$ precipitation accentuates P limitation in CCB as carbonate deposition can trap phosphate in inorganic and organic forms in the growing matrix. A series of field experiments tested these ideas. In the first, investigators determined the balance of calcification on oncoid stromatolites with and without snails. In the next series of experiments, oncoid stromatolites were fertilized with P, and responses of biomass C:N:P ratios, microbial communities, and performance of snails were assessed. Stromatolite C:P and N:P ratios were high but decreased under P fertilization. Surprisingly, P fertilization stimulated snail performance when C:P was reduced moderately but inhibited growth and increased mortality when C:P was strongly reduced, leading us to propose a "stoichiometric knife edge" for snails, with both high and low C:P being detrimental. Finally, studies in Rio Mesquites explored the process of calcification and sup-

J. Corman
School of Life Sciences, Arizona State University, Tempe, AZ, USA

School of Natural Resources, University of Nebraska, Lincoln, NE, USA

J. Elser (✉)
School of Life Sciences, Arizona State University, Tempe, AZ, USA

Flathead Lake Biological Station, University of Montana, Polson, MT, USA
e-mail: Jim.elser@umontana.edu

© Springer International Publishing AG, part of Springer Nature 2018
F. García-Oliva et al. (eds.), *Ecosystem Ecology and Geochemistry of Cuatro Cienegas*, Cuatro Ciénegas Basin: An Endangered Hyperdiverse Oasis, https://doi.org/10.1007/978-3-319-95855-2_5

55

ported the concept of an active "competition" for P between calcification and biotic P uptake. We emerge with a view that CCB's intense P limitation is at least in part a result of widespread CaCO₃ precipitation and that this P limitation imposes a strong stoichiometric limitation on herbivore growth and thus is likely a strong selective factor for consumers.

Keywords Carbonates · C:N:P ratios · Phosphorus · Rio Mesquites · Stromatolites

Introduction

Ecological stoichiometry (ES) is the study of the balance of energy and multiple chemical elements (especially C, N, and P) in ecological interactions (Sterner and Elser 2002). Originally formulated for study of plankton food webs, the ES approach has been extended to diverse biota and ecosystems, including the stromatolite-based food webs of CCB (Elser et al. 2005a, b, 2006). One of the key focus of ES is on the effects of stoichiometric food quality; that is, what happens when a relatively P-rich, low C:P consumer (e.g., a crustacean or a snail) consumes P-limited food (e.g., algae) with high C:P ratio? Abundant data from *Daphnia* (planktonic "water fleas") studies in the lab and in the field have established that high C:P algae indeed result in reduced growth and reproduction due to insufficient P intake to construct P-rich tissues (Hessen et al. 2013). Similar findings have been obtained for aquatic and terrestrial insects (Frost and Elser 2002; Perkins et al. 2004). Thus, high C:P food serves as a kind of "junk food" in the trophic system, reducing ecological transfer efficiency and reducing consumer biomass. More recently, attention has shifted to the other end of the stoichiometric spectrum: P-rich food with low C:P ratio that is also stoichiometrically imbalanced (too P-rich) with respect to the consumer's needs. While data are considerably more sparse, it also appears that food that is stoichiometrically imbalanced on the low C:P side of the spectrum can also lead to reduced consumer performance (Boersma and Elser 2006; Currier and Elser 2017). CCB offers an exciting opportunity to test the effects of stoichiometric imbalance given the widespread prevalence of P limitation in the basin. That is, nutrient concentrations (N and P) are not only low overall, but they are also stoichiometrically imbalanced with ratios of total N to total P commonly >75:1 (by moles). Note that strong P limitation is thought to become established when N:P exceeds ~30 (Downing and McCauley 1992) and that "balanced" N:P ratios in the ocean are ~16 (Redfield 1958). As a result, the biomass that makes up the base of the food web in many CCB systems, "microbialites" (see the following), can often extremely high C:P ratios, often exceeding 1000:1 or even 2000:1 (by moles; see below).

Microbial communities in Cuatro Ciénegas sometimes produce astounding laminated, carbonate structures. These microbialites exhibit a variety of shapes, from small, highly dendritic golf ball-sized spheres in El Mojarral to oblong footballs in Rio Mesquites to platform structures in the Pozas Azules (Fig. 5.1). The microbialites of CCB are thought to be modern analogues of ancient life forms that are found

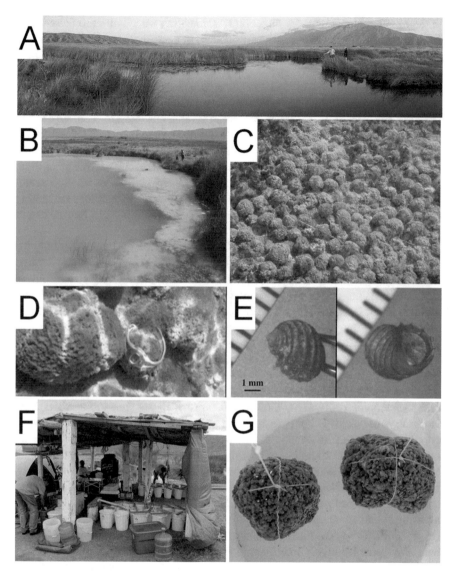

Fig. 5.1 (**a**) The Rio Mesquites. (**b**) Poza Azul with fringing bench of stromatolite reef. (**c**) Oncolite stromatolites littering the bed of Rio Mesquites. Each oncoid is approximately 5–10 cm in diameter. (**d**) Elsewhere in Rio Mesquites, oncoids can reach 20–30 cm in diameter; see watch for scale. (**e**) Typical individual of *Mexithauma quadripaludium* (dorsal and ventral view). (**f**) The experimental setup on the banks of the Rio Mesquites where effects of P on stromatolites and grazing snails were studied (Elser et al. 2005a, b, 2006). (**g**) The interior of an experimental unit seen in (**f**) showing two oncoid stromatolites suspended in Rio Mesquites water. Individually marked snails can be seen (colored dots on the surface of each stromatolite). (Photo credits: J. Elser (**a–c**, **e–g**); J. Corman (**d**))

in the early fossil record as "stromatolites." (Hereafter, we will refer to these living formations at CCB as "stromatolites" although, in a technical sense, that term pertains only to the fossil remains.) Microbial metabolism is the main process by which these "oncoid" (spherical or egg-shaped) stromatolites (sometimes referred to as "oncolites") are formed (Garcia-Pichel et al. 2004). As the photoautotrophic members of the community photosynthesize, they remove dissolved inorganic carbon from the water, tending to increase pH. More alkaline pH tends to promote CaCO3 ($CaCO_3$) deposition, while more acidic pH, resulting from increased aerobic respiration, may cause $CaCO_3$ dissolution (Stumm and Morgan 2012). The most important microbes responsible for $CaCO_3$ deposition in Rio Mesquites are the green alga, *Gongrosira calcifera* Krieger, the cyanobacteria *Homoeothrix balearica* Bornet and Flahult and *Schizothrix lacustris* A. Brown, and a diverse assemblage of benthic diatoms (Winsborough 1990). Quite simply, these stromatolites can be considered "living rocks built by algae" (Dinger et al. 2006).

The microbialite-based food webs of CCB are particularly exciting to study for a variety of reasons, one of which is that they provide a type of ecological "time machine" that allows study of Earth's first food chains. Such "living stromatolites" have long been thought to be rare on the modern Earth (the most famous ones being found at Shark Bay, Australia), but diverse manifestations of such extant communities are found throughout CCB, including in the Rio Mesquites (see below) and in the spectacular El Mojarral ponds in the southeast part of the basin (Fig. 5.1). Another interesting feature of CCB's stromatolites is that they support a food chain. In particular, they are grazed upon by endemic hydrobiid snail *Mexithauma quadripaludium* and to a lesser extent by another endemic hydrobiid snail, endemic hydrobiid *Nymphophilus minckleyi*. This is in contrast with the situation at Shark Bay, where the absence of animal grazing (due to the bay's high salinity) is thought to allow persistence of its famous stromatolites. This begs the question about what may constrain the proliferation of grazing animals in Cuatro Ciénegas habitats and led us to the hypothesis that strong ecosystem P limitation results in high C:P ratios in stromatolite biomass, thus imposing strong stoichiometric food quality constraints on snails and other consumers, keeping them at low levels at CCB, and allowing abundant and diverse stromatolite communities to proliferate and persist (Elser 2003). This is the main hypothesis tested in the set of experiments described here. We were also highly interested in finding out if the intensity of calcification (formation of $CaCO_3$ precipitates) at CCB contributed to ecosystem P limitation and thus to regulating food web dynamics in these habitats.

Experiments/Studies

Study site: The Cuatro Ciénegas Basin (CCB) lies in a Cretaceous karst landscape, with springs draining rocks made up of shallow marine limestones, dolomite, and gypsum. The springs and spring-fed streams and ponds are rich in minerals, in particular: sulfates, carbonates, calcium, magnesium, and sodium

(Johannesson et al. 2004). The main flow system is the "Garabatal-Becerra-Rio Mesquites" system in the northern part of the valley, fed by the Cupido-Aurora aquifer. Part of this flow system, Rio Mesquites, is the largest stream in CCB (Fig. 5.1a). It begins with headwaters at Poza Mojarral Este and meanders east through the desert, filling various wetlands and ponds and, ultimately, into wetlands in the eastern side of the valley associated with a large, shallow lake known as Las Playitas. The springs that feed these very clear waters remain near 30–35 °C all year. The waters of Rio Mesquites, like much of CCB valley, are oligotrophic. Mean nitrogen (N) concentrations range from 0.82 to 0.87 mg N/L, and mean total phosphorus (P) concentrations are quite low, ranging from 15 to 19 µg P/L (as total P; Elser et al. 2005a, b).

Experiments: A key piece to understanding the presence of stromatolites in CCB is to describe how they can form and grow despite the presence of metazoan grazers. Garcia-Pichel et al. (2004) studied this paradox by comparing rates of calcification to those of bioerosion. Of course, for stromatolites to persist in Cuatro Ciénegas, they must calcify at a higher rate than they are eroded (or dissolved). Yet, prior to the Garcia-Pichel et al. (2004) study, this had not been studied in CCB or any other location with living stromatolites. The authors collected stromatolites from Rio Mesquites and incubated them in buckets in a laboratory with water from Rio Mesquites. They measured calcification rates (under different light conditions) by determining mass transfer rates of Ca^{2+} and oxygen through the stromatolite boundary layer and based on changes of bulk Ca^{2+} concentration in the buckets. The authors measured rates of snail bioerosion by monitoring fecal pellet production; snail fecal pellets consisted of ~74% carbonates. They also found that fecal pellet production, and, hence, stromatolite bioerosion, was a direct linear function of the number of snails present on a stromatolite. Extrapolating their results spatially, they found gross calcification rates of 1.49 ± 0.12 kg $CaCO_3$ m^{-2} y^{-1} and bioerosion rates of 1.13 ± 0.1 kg $CaCO_3$ m^{-2} y^{-1}. Hence, snail grazing potentially accounted for a loss of nearly 75% of the mineralized growth of the oncoids (Garcia-Pichel et al. 2004).

Three primary experiments in the Rio Mesquites followed up on this preliminary study to investigate the interactions among phosphorus, stromatolites, and snails based on a hypothesis that low P availability for stromatolite microbes meant that they would have high C:P ratio in their biomass, imposing a stoichiometric limitation on herbivorous snails. Experiments 1 and 2 are reported in Elser et al. (2005a, b), while Experiment 3 is reported in Elser et al. (2006). In each of these experiments, one or two individual oncoids were suspended in 20-L buckets containing stream water on the shores of the stream (Fig. 5.1f). Water was replenished 2–3 times daily and circulated intermittently using submersible pumps. Prior to each experiment, all snails were manually removed from each stromatolite, and then a standard number of snails (equivalent to ambient densities) were returned to each. Stromatolites were suspended ~5 cm above the bottom of each bucket so that the snails could not leave (Fig. 5.1g). *Experiment 1* in summer 2001 simply tested the effects of P enrichment on Rio Mesquites' oncoid stromatolites and their associated snails. Then, stromatolites were randomly assigned to two treatments: control (no enrichment) and P enrichment (+15 μmol/L P). Enrichment occurred every time bucket water was replenished. This experiment ran for 3 weeks. *Experiment 2* (summer 2002) involved a factorial manipulation of P enrichment and presence/absence of snails: −P,−S (no enrichment, no added snails); −P,+S (no enrichment, snails added back the stromatolite at approximately ambient density); +P,−S (P enrichment by 5 μmol/L P, no snails), and +P,+S (P enrichment, snails added). The experiment ran for 7 weeks. *Experiment 3* involved a similar approach, but instead of a single level of P enrichment, several levels (0, 0.1, 0.25, 0.5, 1.25, 2.5, 5, and 10 μM) were applied. This experiment ran for 5.5 weeks during the summer of 2003.

At the end of each experiment, a variety of chemical and biological measurements were made, including assessment of stromatolite microbial community structure (via DGGE and TRFLP), process rates (primary production and respiration, calcification, P removal), and biomass (organic matter content) and C:P ratio of surficial layers (note that our methods removed C from carbonates prior to analysis). Effects on snails were also assessed by quantifying body size, survival, RNA/DNA ratios (as an index of growth rate), and growth rate (via changes in operculum size for marked individuals; Fig. 5.1e). Effects of P enrichment on the stromatolites were consistent: while generally changes in biomass levels (e.g., organic matter content) could not be detected, P enrichment consistently altered microbial community structure (lowering diversity) and stimulated rates of production, respiration, and calcification. P enrichment also lowered the C:P ratio of the organic matter on the stromatolite surface and did so in proportion to the level of P added. In Experiment 1, C:P ratio declined to ~500:1 (molar) from very high levels of ~2300:1 (Fig. 5.2a). In Experiment 2, stromatolite C:P ratios were somewhat lower in general at the time of the experiment, and thus P enrichment lowered C:P from ~750:1 to ~80:1 (Fig. 5.2b). Finally, in Experiment 3, adding P at increasing levels lowered C:P from ~1200:1 to ~130:1 (Fig. 5.2c).

Effects of P enrichment on snail performance were strong and, initially at least, confusing (Fig. 5.2). In Experiment 1, snails showed a strong body-size dependence of body P content (% dry mass) and RNA/DNA ratio, consistent with the likelihood that smaller snails have faster growth rates than larger ones. Furthermore, snails in the P enrichment treatment has significantly higher %P and RNA/DNA ratio (Fig. 5.2a) than those in the control treatment. This supports the hypothesis that low stromatolite P content (high C:P ratio) limits snail growth. When the experiment was repeated the following year, a different result was obtained. While snails exhibited a consistent body-size decline in P content, RNA/DNA ratio, and growth rate (measured directly by following marked individuals in this experiment), in this experiment P enrichment led to *lower* P content, RNA/DNA ratio, and growth rate (Fig. 5.2b) than in controls. Furthermore, mortality rates of snails increased from ~6% in controls to ~37% in the +P treatment. In interpreting these divergent results, Elser et al. (2005b) proposed that snails live on a "stoichiometric knife edge," in which enrichment of high C:P stromatolite biomass in Experiment 1 stimulated growth by reducing stoichiometric imbalance, but in Experiment 2, stromatolite biomass had moderate C:P ratios initially and then,

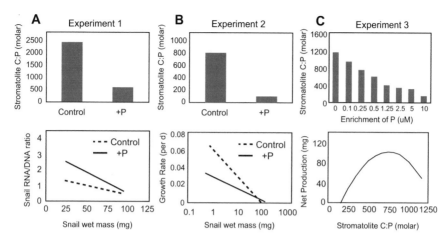

Fig. 5.2 Effects of P enrichment on stromatolite C:P ratio and subsequent effects on snail performance. (**a**) Experiment 1 in which P enrichment lowered stromatolite C:P (molar) from very high values (>2000) to moderate levels (~400) and stimulated snail RNA/DNA ratio. (**b**) Experiment 2 in which P enrichment lowered stromatolite C:P from moderate levels (~750) to low levels (<100) and lowered snail growth rate. (**c**) Experiment 3 in which different levels of P enrichment lowered stromatolite C:P moderately high levels (1200) to low levels (130), leading to a response of total snail production (combination of effects on growth and survivorship) with a peak at intermediate stromatolite C:P, a "stoichiometric knife edge"

after enrichment, achieved very low values that were excessively P-rich for snails, leading to reduced growth rate and increased mortality. The existence of this "knife edge" response was confirmed in several other studies involving insects, crustaceans, mollusks, and fish (Boersma and Elser 2006). Experiment 3, which involved a gradient of P enrichment, was designed to test the existence of this hump-shaped function (Elser et al. 2006), and indeed its existence was confirmed. Low to moderate levels of P enrichment led to stimulation of growth rates of small snails, but high levels of P enrichment reduced snail growth rates. Similarly, snail survivorship was highest at intermediate P enrichment levels. When growth and survivorship responses were combined into a single population-level product metric, production showed a very strong hump-shaped relationship with stromatolite C:P ratio (Fig. 5.2c). These results led Elser et al. (2006) to propose a P-centric model of early animal evolution when metazoans first proliferated in stromatolite-dominated shallow seas during the Precambrian/Cambrian transition during a time that Earth's P cycle became very dynamic (as evidenced by the geologic phosphorite deposition record). In this model, emergence of metazoans was constrained by P limitation during Earth's early history, prior to large-scale P weathering on continents. Later, enrichment of the oceans by elevated P coming from the continents stimulated proliferation and diversification of metazoans (primarily Ediacaran fauna) and, in some cases, led to excesses of P (the negative side of the "knife edge") that selected for mechanisms to sequester excess P (e.g., formation of biological Ca-phosphate apatite minerals) that eventually gave rise to bone-forming vertebrates (the "Cambrian explosion").

The experiments just reviewed support the hypothesis that the food webs on stromatolites in Rio Mesquites are consistently phosphorus-limited, begging the question: why is P so limiting to biological production in Cuatro Ciénegas? Corman et al. (2016a, b) set out to test one possible explanation: interactions between $CaCO_3$ and phosphorus. When $CaCO_3$ forms, it can adsorb or coprecipitate P in the form of phosphate (House et al. 1986). Hence, that phosphorus may be no longer biologically available and limit primary production (Corman et al. 2016a, b). In the first experiment, the resource-limitation experiment, small stromatolites (~17 cm³) were placed in 2-L plastic containers containing stream water from Rio Mesquites. Containers were incubated in the bottom of the stream and water was replaced twice daily. This 16-day experiment tested the effects of supplemental dissolved organic carbon and P additions on stromatolite metabolic activity (photosynthesis and respiration) and calcification. In the second experiment, the calcification experiment, larger stromatolites (~102 cm³) were placed in 6-L plastic containers containing stream water from Rio Mesquites. Containers were incubated off-site, but stream water was replenished every other day, and water flow, temperature, and light conditions roughly mimicked those conditions in Rio Mesquites. This 5-week experiment tested the effects of altered calcification rates, achieved by either removing light, decreasing Ca^{2+} concentrations, or adding a calcification inhibitor, strontium, on the stromatolite ecosystem. In a third experiment, we tested nutrient limitation of photoautotrophic microbes not associated with stromatolites. Agar-filled jars amended with either N, P, or both ("nutrient-diffusing substrates") were incubated in Rio mesquites. After a couple of weeks, chlorophyll *a* abundance, a proxy of photoautotroph biomass, was compared across the treatments and control jars.

Results from the resource-limitation experiment supported earlier work by Elser et al. (2005a): enrichment with P increased primary production and respiration (Fig. 5.3a). Results from the calcification experiment supported the role of calcification in limiting P availability: when calcification rates were lowered through Ca^{2+} removal, organic C and P accumulation increased in the microbial biomass. In the first 3 weeks, P accumulation in the Ca^{2+} removal treatment increased 121% (Fig. 5.3c). In the calcification experiment, we also found that the light removal treatment leads to a drastic change in the stromatolites; microbial biomass quickly deteriorated and stromatolites turned gray. There was no or slightly negative organic carbon accumulation in the first 3 weeks, suggesting photoautotrophs were no longer growing. The phosphorus concentration in the biomass also decreased in the light removal treatment. Therefore, while calcification may be lowering P availability, photoautotrophs likely play an important role in cycling and retaining whatever P is available in these stromatolites. Additionally, we found that photoautotrophs not associated with stromatolites, i.e., those growing on the nutrient-diffusing substrates, were most likely to respond to concomitant additions of N and P. Hence, photoautotrophs growing on stromatolites are distinctly P-limited (Fig. 5.3b).

These experiments also shed new light on the process of calcification in stromatolites. Unlike the studies discussed above (Elser et al. 2005a), P enrichment did not stimulate calcification. Instead, the organic C enrichment stimulated calcification (along with dampening the autotrophic effects of P on primary production and stimulating aerobic respiration). While this result was unexpected, it supports

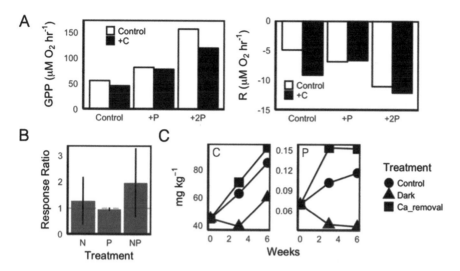

Fig. 5.3 Interactions between phosphorus availability, microbial metabolism, and calcification in stromatolites. (a) The resource-limitation experiment in which increases in gross primary production (GPP) were damped by increases in organic carbon (C) and increases in respiration (R) occurred with increases in both P and C. (b) The nutrient-diffusing substrate experiment in which photoautotrophs not associated with stromatolites do not exhibit strong P limitation. (c) The calcification experiment in which decreases in calcification rate through calcium removal (Ca removal) lead to increased C and P accumulation in stromatolites

conclusions from genomic analysis of stromatolites that heterotrophic processes, e.g., sulfate reduction, may at times contribute to calcification (Breitbart et al. 2009; Nitti et al. 2012).

Implications/Conclusions

Overall, these experiments support the view that food webs at Cuatro Ciénegas are strongly constrained by limiting supplies of P. Relative supplies of N and P are very strongly imbalanced; N:P ratios of total nutrient pools in the water column of Cuatro Ciénegas are among the highest observed. Coupled to this stoichiometric imbalance in nutrient supplies is strong imbalance among C, N, and P in stromatolite biomass. Accordingly, the C:P ratios and N:P ratios of surficial biomass of the stromatolites are also highly imbalanced. Indeed, the C:P values seen in Experiment 1 (~2200:1 by moles) are among the highest recorded for aquatic biota and are likely matched only by highly woody (cellulose-rich) tissues of terrestrial plants. These C:P ratios are highly responsive to P supply as seen in all of three of the experiments reviewed. It is also notable that stromatolite C:P ratio varies strongly over time. This is seen in the C:P ratio of unenriched stromatolites in the three studies (Fig. 5.2): ~2200:1 in Experiment 1, ~750:1 in Experiment 2, and ~1200:1 in Experiment 3. Thus, stromatolite grazers must adapt to food quality conditions that are highly variable across years. We have little idea about the causes of this year-to-year variation nor do we know the extent of intra-annual variation.

One hint about year-to-year variation in stromatolite C:P ratios comes from the observation of the relationship between P enrichment and calcification. In the experiments by Elser et al., the relationship was stimulatory. In the experiment by Corman et al., the relationship was not detected. Hence, the relative abundance or metabolic activity of different microbes in the microbial community may be changing temporally and, ultimately, be reflected in differing C:P ratios of the combined biomass.

Our finding of both positive (stimulatory) and negative (inhibitory, perhaps toxic) effects of P enrichment on stromatolite-grazing snails has both short-term and long-term implications. In the short-term, the negative effects of P enrichment highlight the importance of protecting the P-limited ecosystems from external P inputs, either from human or animal waste or from agricultural runoff or atmospheric transport of P. Such inputs would not only disrupt the structure and function of the stromatolites themselves; if sustained at low levels, over time enrichment might lead to proliferation of grazers and bioerosion of the stromatolites. If P inputs occurred episodically but at high levels, stromatolite C:P ratios might be pushed "over the hump," to levels low enough to cause reduced growth and increased mortality of snails. This is of some concern as Cuatro Ciénegas snails are endemic and thus part of the basin's highly valued biodiversity. From a (very) long-term perspective, our findings of a "stoichiometric knife edge" for this stromatolite-grazer system support at least the feasibility of the scenario sketched by Elser et al. (2006), which argues for a role of P limitation in constraining animal evolution/emergence during early periods of Earth history when the global P cycle was poorly developed and for P excess in driving early evolution of proto-vertebrates that may have sequestered excess P into mineral forms that later provided structural functions.

The relationship between calcification and P bioavailability has interesting implications, as well. Could P limitation in CCB be so severe because of $CaCO_3$-phosphorus interactions? Our results support this notion. Could the stromatolites that once covered the Earth in the Precambrian have dealt with extreme constraints on P acquisition due to calcification, as well? Boring into rocks may be an adaptation by microbes that live on carbonates to deal with this constraint (Cockell and Herrera 2008); it is unknown if microbes living in the CCB also do this but deserves further investigation. If P limitation in the CCB is so severe because of calcification and coprecipitation of phosphate, then processes that control calcification rates are indirectly also very important for regulating P availability. If calcification rates decrease or if environmental conditions are no longer thermodynamically favorable for calcification, P supplies may increase. Decreases or losses of calcification could occur if pH or alkalinity decreases (Stumm and Morgan 2012), as is occurring in the oceans with increased atmospheric CO_2. While the relationship between atmospheric CO_2 partial pressure and pH in inland waters is more complex, recent investigations suggest alkaline lakes may be susceptible to acidification (Weiss et al. 2018). If the same feedbacks happen in CCB, P supplies may increase with profound implications for stromatolite food webs.

References

Boersma M, Elser JJ (2006) Too much of a good thing: on stoichiometrically balanced diets and maximal growth. Ecology 87:1325–1330

Breitbart M, Hoare A, Nitti A et al (2009) Metagenomic and stable isotopic analyses of modern freshwater microbialites in Cuatro Ciénegas, Mexico. Environ Microbiol 11:16–34

Cockell CS, Herrera A (2008) Why are some microorganisms boring? Trends Microbiol 16:101–106

Corman JR, Moody EK, Elser JJ (2016a) Calcium carbonate deposition drives nutrient cycling in a calcareous headwater stream. Ecol Monogr 86:448–461

Corman JR, Poret-Peterson AT, Uchitel A et al (2016b) Interaction between lithification and resource availability in the microbialites of Rio Mesquites, Cuatro Ciénegas, México. Geobiology 14:176–189

Currier CM, Elser JJ (2017) Beyond monoculture stoichiometry studies: assessing growth, respiration, and feeding responses of three Daphnia species to P-enriched, low C:P lake seston. Inland Waters 7:348–357

Dinger EC, Hendrickson DA, Winsborough BM et al (2006) Role of fish in structuring invertebrates on stromatolites in Cuatro Ciénegas, México. Hydrobiologia 563:407–420

Downing JA, McCauley E (1992) The nitrogen: phosphorus relationship in lakes. Limnol Oceanogr 37:936–945

Elser JJ (2003) Biological stoichiometry: a theoretical framework connecting ecosystem ecology, evolution, and biochemistry for application in astrobiology. Int J Astrobiol 2:185–193

Elser JJ, Schampel JH, Garcia-Pichel FE et al (2005a) Effects of phosphorus enrichment and grazing snails on modern stromatolitic microbial communities. Freshw Biol 50:1808–1825

Elser JJ, Schampel JH, Kyle M et al (2005b) Response of grazing snails to phosphorus enrichment of modern stromatolitic microbial communities. Freshw Biol 50:1826–1835

Elser JJ, Watts J, Schampel JH et al (2006) Early Cambrian food webs on a trophic knife-edge? A hypothesis and preliminary data from a modern stromatolite-based ecosystem. Ecol Lett 9:295–303

Frost PC, Elser JJ (2002) Growth responses of littoral mayflies to the phosphorus content of their food. Ecol Lett 5:232–240

Garcia-Pichel F, Al-Hourani FA, Farmer JD et al (2004) Balance between microbial calcification and metazoan bioerosion in modern stromatolitic oncolites. Geobiology 2:49–57

Hessen DO, Elser JJ, Sterner RW et al (2013) Ecological stoichiometry: an elementary approach using basic principles. Limnol Oceanogr 58:2219–2236

House WA, Casey H, Donaldson L, Smith S (1986) Factors affecting the coprecipitation of inorganic phosphate with calcite in hardwaters. l. War Res 20:917–922

Johannesson KH, Cortés A, Kilroy KC (2004) Reconnaissance isotopic and hydrochemical study of Cuatro Ciénegas groundwater, Coahuila, México. J South Am Earth Sci 17:171–180

Nitti A, Daniels CA, Siefert J et al (2012) Spatially resolved genomic, stable isotopic, and lipid analyses of a modern freshwater microbialite from Cuatro Ciénegas, Mexico. Astrobiology 12:685–698

Perkins MC, Woods HA, Harrison JF et al (2004) Dietary phosphorus affects the growth of larval *Manduca sexta*. Arch Insect Biochem Physiol 55:153–168

Redfield AC (1958) The biological control of chemical factors in the environment. Am Sci 46:205–221

Sterner RW, Elser JJ (2002) Ecological stoichiometry: the biology of elements from molecules to the biosphere. Princeton University Press, Princeton, NJ

Stumm W, Morgan JJ (2012) Aquatic chemistry: chemical equilibria and rates in natural waters. Wiley, Hoboken, NJ

Weiss LC, Potter L, Steiger A et al (2018) Rising pCO_2 in freshwater ecosystems has the potential to negatively affect predator-induced defenses in Daphnia. Curr Biol 28:327–332.e3

Winsborough BM (1990) Some ecological aspects of modern freshwater stromatolites in lakes and streams of the Cuatro Ciénegas Basin, Coahuila, Mexico. Dissertation, The University of Texas at Austin

Chapter 6
The Sulfur Cycle as the Gear of the "Clock of Life": The Point of Convergence Between Geological and Genomic Data in the Cuatro Cienegas Basin

Valerie De Anda, Icoquih Zapata-Peñasco, Luis E. Eguiarte, and Valeria Souza

Contents

Abstract Due to its chemical properties and several stable redox states, microbial transformations of sulfur compounds have been affecting the geochemical features of the Earth's biosphere since the Archaean. However, despite the great importance of sulfur cycling, reconciling the geologic record with genomic data has been challenging. Here we first review current state-of-the-art evidence about the emergence of life on Earth in sulfur-rich environments, providing a conceptual framework that closely connects these two largely separated disciplines. Then, we summarize the current astonishing diversity of prokaryotes responsible for driving the sulfur cycle, suggesting that, due to their ancient origin, sulfur-associated taxa perhaps hold the greatest diversity of any group of microorganisms to metabolize a single element on Earth. Finally, because the guilds of sulfur metabolizing microbes co-occur at

V. De Anda · L. E. Eguiarte · V. Souza (✉)
Departamento de Ecología Evolutiva, Instituto de Ecología, Universidad Nacional Autónoma de México, Coyoacan, Mexico
e-mail: souza@unam.mx

I. Zapata-Peñasco
Instituto Mexicano del Petróleo, Mexico City, Mexico

© Springer International Publishing AG, part of Springer Nature 2018   67
F. García-Oliva et al. (eds.), *Ecosystem Ecology and Geochemistry of Cuatro Cienegas*, Cuatro Ciénegas Basin: An Endangered Hyperdiverse Oasis,
https://doi.org/10.1007/978-3-319-95855-2_6

millimeter scales within microbial mats, we use the taxonomic and metabolic information derived from these primordial communities as ecological models to highlight sulfur as the guiding axis of these complex intersections, recapitulating a gear of the clock of life.

Keywords Bacteria · Diversity · Microbial mat · Prokaryotes · Sulfur metabolism

Oldest Evidence of Life on Earth: Stromatolites, Hydrothermal Vents, and Hot Springs – What Do They Have in Common?

Since its discovery in the early 1980s, scientific consensus indicated that the Dresser Formation, Pilbara Craton, Western Australia, contained evidence of the earliest forms of life on Earth. Paleobiological and chemical data suggest that the stromatolites within the Dresser Formation (a region underlain by a 30-km thick sequence of relatively well-preserved sedimentary and volcanic rocks) are approximately 3,465 million years old (Ma) and hold the oldest and most convincing signs of life in the form of domical and coniform layered stromatolites (Allwood et al. 2006; Van Kranendonk et al. 2008). The biogenicity of these fossils was carefully confirmed by morphology, stable isotope signatures, seawater-like trace element signatures, and the presence of microfossils that were identified as the earliest known assemblage of cellular fossils (Allwood et al. 2006; Noffke et al. 2013; Schopf 1993; Tice and Lowe 2004; Van Kranendonk et al. 2008; Walter et al. 1980; Westall et al. 2006). More than two decades ago, this assemblage of microfossils was associated with the morphological taxa of prokaryotic, filamentous, and coccoidal microorganisms (Schopf 1993) and was recently validated using secondary ion mass spectroscopy. Other interpretations of the Dresser Formation, using sulfur isotopes, indicate the early existence of organisms that disproportinate elemental sulfur (Philippot et al. 2007). Evidence of microbe-driven sulfate reduction has also been found in sedimentary settings (Shen et al. 2001, 2009; Ueno et al. 2008) and hydrothermal seafloor samples (Aoyama and Ueno 2018; Shen et al. 2001, 2009; Ueno et al. 2008). The presence of these prokaryotic assemblages, consisting of photosynthesizers, archaeal methane producers, methane consumers, and sulfur users (Schopf et al. 2018), as well as the diversity of the stromatolites, points to a biochemical sophistication of life by that time, 3,500 Ma ago, in which the biosphere had already given rise to an ancient but metabolically diverse ecosystem. Consequently, life must have originated significantly earlier on our planet. More recently, evidence points toward the limits between the Hedean and the Archaean period, 4.1 billion years ago, where the life seems to have left an isotopic imprint in two independent locations (Dodd et al. 2017; Tashiro et al. 2017).

Recent evidence has confirmed that in fact life was present even hundreds of millions of years earlier than even the Dresser Formation. First, recently discovered stromatolites from the Isua supracrustal belt (ISB) in southwest Greenland (Nutman et al. 2016) are dated at 3,700–3,800 Ma years old and indicate macroscopic layered structures produced by microbial communities that, according to trace elements signatures, grew in a shallow marine environment (Dalton 2004). Second, recent hypotheses point out that thermal sulfur-related environments, such as hot springs or hydrothermal vents, where high temperatures might have increased the rate of primordial biochemical enzymatic reactions and where high concentrations of reducing equivalents made possible chemiosmosis gradients across a cell membrane, could indeed have been the right places for life to emerge (Dai 2017; Djokic et al. 2017; Van Kranendonk et al. 2017). Clear evidence of this has been found in precipitates from the Nuvvuagittuq belt in Quebec, Canada, interpreted as putative fossilized microorganisms in ferruginous sedimentary rocks deduced as seafloor hydrothermal vent related that are at least from 3,770 and possibly 4,280 Ma (Dodd et al. 2017). Other fossil evidence, though not as ancient as the Quebec precipitates, points to an active volcanic landscape, in contrast to deep hydrothermal settings. Stratigraphic and geochemical evidence indicates that recently discovered Archaean hot spring deposits of 3,500 Ma, found also in the Dresser Formation, were not so different from those present today and were affected by voluminous hydrothermal fluid circulation preserving a suite of microbial bio-signatures (Djokic et al. 2017; Van Kranendonk et al. 2017).

As can be seen, the recent evidence from geological record has shown that the right physical and chemical settings for the origin of life can be found in hot springs and hydrothermal vents, where the oldest evidence is possibly 4,280 Ma. Under this scenario, the theory of the iron-sulfur world suggests that life started chemo-autotrophically in S-rich environments, such as aqueous flows of volcanic outflows where the oxidative formation of pyrite satisfies all the necessary conditions needed by an energy source for such an origin (Wächtershäuser 1990, 2008).

The geologic record suggests that these oldest primordial environments (mats, hot springs, and hydrothermal vents) are the sites where a bridge between sulfur chemistry and life arose through energetically favorable reactions that led to one of the most important biological processes on Earth, "the fixation of carbon dioxide into organic materials" (see Chap. 1). Among these energetically favorable reactions, the most accepted sources of energy available for chemosynthesis in early Earth were hydrogen gas (reacting with oxidants such as carbon dioxide) and sulfur dioxide (because, under anaerobic conditions, the reaction of CO_2 with H_2 releases energy). Current modern microorganisms using these ancient forms of chemical energy (H_2/CO_2 and H_2/H_2S couples) are found in sulfur-rich deep environments such as hydrothermal vents of the seafloor (Staley 2002). In fact, several lines of evidence support sulfide (H_2S) over H_2 as the primordial energy source (reviewed in detail in Olson et al. (2016)). For example, due to its chemical properties, sulfide can serve as an important organic product, reactant, catalyst (proto-enzyme), barrier (proto-membrane), and sustainable source of energy (Olson et al. 2016). Furthermore, recent studies suggest a critical role of FeS-dependent enzymes and of thioesters as carriers of chemical energy in the assembly of early metabolic networks. In fact, the exergonic reaction to form a thioester is the basis of the most ancient form of CO_2 fixation, the acetyl-CoA pathway (Semenov et al. 2016; Goldford et al. 2017; Martin and Thauer 2017).

The paleontological and geological record depict the environment where the evolution of life occurred, and based on the assumption that molecular sequence determines biological function and that the ancient molecules would have necessarily been adapted to function in their surroundings, well-conserved molecules should provide evidence of past environmental conditions (Garcia et al. 2017). Thus, the molecular reconstruction of life using completely sequenced genomes (phylogenomics) can give evidence about the dominant conditions in the paleoenvironment under study (Garcia et al. 2017).

Reconciliation with standard models of microbial evolution derived from phylogenies indicates that early branches are occupied by thermophilic and hyperthermophilic prokaryotes capable of a wide range of S oxidation and reduction processes. This suggests that, in early ecosystems, microbial communities exploited the full range of the S-redox spectrum, with some organisms reducing and others oxidizing the reduced sulfur species. Hence, microbes from the "FeS world" were already building diverse stromatolites before the origin of cyanobacteria (Grassineau et al. 2001). In fact, the grouping of hyperthermophilic species on short branches near the bases of both the bacterial and archaeal domains of the tree parsimoniously suggests that LUCA (last universal common ancestor) was a hyperthermophile (Lake et al. 2009; Gaucher et al. 2010). Moreover, respiration of sulfur compounds is thought to be one of the first metabolisms to evolve in a hot primitive Earth. Indeed, the placement of hyperthermophiles Archaea and deep-branching thermophilic bacteria is consistent with such an early origin. Isotopic data also suggest that sulfate reduction began around 3 billion years ago, and acquired global significance only after sulfate concentrations had significantly increased in the Precambrian oceans (Trudinger 1992; Ueno et al. 2008; Zhelezinskaia et al. 2014). The conservation of the sulfate reduction metabolism (using dissimilatory sulfite reductase) in both prokaryotic domains of life (Archaea and Bacteria) suggests two main hypotheses. The first one suggests that this trait could be present in the last common ancestor before the divergence of the two domains. The second hypothesis suggests that such a metabolic trait could have evolved in one of the domains soon after the divergence and was then transferred to the other domain by several early lateral gene transfer events. Given the expected parsimony of evolution, deep ancestry before divergence is the most accepted hypothesis (Wagner et al. 1998; Muyzer and Stams 2008; Müller et al. 2015).

Using Sulfur in Early Life: Molecular and Genomic Evidence

In agreement with the geological record, the metabolic implications of so-called "standard" rRNA models suggest that the full sulfur cycle predated oxygenic photosynthesis and thus was in operation before cyanobacteria had evolved and were building diverse stromatolites (Grassineau et al. 2001; Staley 2002). Recent evidence suggests that LUCA (last universal common ancestor) was a hyperthermophile (Lake et al. 2009; Gaucher et al. 2010) and was confined to FeS compartments geologically produced at vents on the ocean floor that provided the initial cell "wall" (Dai 2017). Compartmentalization is a prerequisite for the evolution of any complex system, and it seems most possible that this configuration consisted of iron sulfide (FeS) deposited at a warm (<90 °C) submarine hydrothermal spring (Koonin and Martin 2005).

Modern Distribution of Sulfur Metabolism Across the Tree of Life

Currently, microbial usage of sulfur compounds is reflected in the astonishing diversity of sulfur-related metabolic pathways that strongly correlate with environmental and ecological influences. The great diversity of sulfur-related microorganisms on Earth can be observed as a result of continual innovations in molecular pathways, which involve large sets of enzymes, organic substrates, and electron carriers, which in turn ultimately depend on particular redox potentials, organic matter availability, electron donor or acceptor disposal, and temperature, among others (Canfield et al. 2005; Ghosh and Dam 2009; Hubas et al. 2011). This complexity increases when considering the surrounding geochemical and ecological conditions of the sulfur players and the syntrophic relationship and shared chemical energy across space and time (Plugge et al. 2011; Ozuolmez et al. 2015; Lau et al. 2016). If you look carefully to the tree of life, you will find that the use of sulfur is widespread along the Bacteria and Archaea domains. Clearly, metabolic types are typically not confined to monophyletic clades. Conversely, species that do not use sulfur as an energy source are often closely related to sulfur users. Historically, the S taxa have been distinguished based on combined considerations derived from physiological, biochemical, and evolutionary traits, e.g., membrane structure, pigment composition, or phylogenetic position, and are grouped together into five well-established metabolic guilds that we briefly describe below:

1. Colorless sulfur bacteria (CLSB)

This guild is historically named for their lack of photopigments, although some are colorful (pink or brown) due to their high cytochrome content. They have been subdivided with respect to their metabolic capabilities into the following four categories. (A) Obligate chemolithotrophs use inorganic reduced sulfur compounds as energy source and use the Calvin cycle for carbon fixation. In the most recent tree of life classification, this subgroup of CLSBs is scattered among the Bacteria and Archaea domains of life, including *Betaproteobacteria* (e.g., *Thiobacillus* and *Thiomonas*), *Gammaproteobacteria* (*Thiomicrospira*), *Aquificae* (e.g., *Hydrogenobacter*), and Crenarchaeota (e.g., *Sulfolobus*). (B) The second class of CLSB is composed of the facultative chemolithotrophs, which can grow on mixtures of reduced sulfur compounds and organic substrates either chemolithoautotrophically or heterotrophically with complex organic compounds used as both carbon and energy sources. In contrast to the first group, the facultative chemolithotrophs are restricted within the *Proteobacteria* group: *Gammaproteobacteria* (e.g., *Thiosphaera*, *Beggiatoa*) and *Alphaproteobacteria* (e.g., *Paracoccus*). (C) The third type, the chemolithoheterotrophs, also gains energy from the oxidation of reduced sulfur compounds, but does not fix carbon dioxide. This metabolism is also found in some species of *Gamma-* and *Betaproteobacteria* (e.g., *Thiobacillus* and *Beggiatoa*, respectively). (D) The last group of CLSB encompasses chemoorganoheterotrophic microorganisms that possess the ability to oxidize reduced sulfur compounds, but do not appear to derive energy from them. These microorganisms are placed within the *Gamma-* and *Betaproteobacteria* classes (e.g., *Thiobacterium*, *Thiothrix*, and *Macromonas*, respectively (Robertson and Kuenen 2006)).

2. Purple sulfur bacteria (PSB)

These are named for their pigmentation due to the presence of bacteriochlorophyll *a* and *b*. These metabolically versatile bacteria can grow photolithoautotrophically using low light, relying on reduced sulfur compounds as electron donors, and fixing carbon through Calvin cycle. The PSB are divided into two families, the *Chromatiaceae* and the *Ectothiorhodospiraceae* within the *Gammaproteobacteria* division.

3. Green sulfur bacteria (GSB)

GSB are also named for their bacteriochlorophyll content, but unlike PSB, the GSB harbors c-, d-, e-, and f-type bacteriochlorophyll. They also use the reductive (reverse) tricarboxylic acid (RTCA) cycle, for carbon dioxide fixation, instead of the Calvin cycle. In addition, most GSB can assimilate a small number of simple, organic compounds, such as acetate, but only in the presence of CO_2 and a photosynthetic electron donor. Most strains use electrons derived from oxidation of sulfide, but some strains can also oxidize elemental sulfur, thiosulfate, hydrogen, and iron(II) (Frigaard and Bryant 2008).

Overall, the oxidative sulfur metabolism in the photosynthetic anoxygenic bacteria shares some chemical and ecological similarities; however, their physiology and evolution are rather different. For example, the GSB are less physiologically versatile compared with the PSB (Friedrich et al. 2001; Frigaard and Bryant 2008).

The metabolism involved in the reduction of sulfur compounds has also been classified based on the ecological, biochemical, and evolutionary traits of the microorganisms involved: sulfate-reducing microorganisms (SRM) vs. elemental sulfur-reducing microorganisms (ESRM). There are also organisms that use other oxidized sulfur compounds, in particular thiosulfate and sulfite, as electron acceptors in their energy metabolism. However, very little is known about obligate thiosulfate or sulfite reducers. Usually, they are subsumed in the SRM class (Canfield et al. 2005).

4. Sulfate-reducing microorganisms (SRM)

These anaerobic microorganisms mainly use sulfate as a terminal electron acceptor in an energy-gaining respiratory process linked to the oxidation of an electron donor. The electron transfer pathway of this metabolism is not well understood (Pereira et al. 2011). The range of substrates used by SRM is very broad (e.g., dicarboxylic acids, alcohols, amino acids, sugars, a wide variety of aromatic compounds, benzene, straight chain alkanes, and inclusive man-made xenobiotic compounds and hydrogen) (Canfield et al. 2005). Such a large substrate amplitude suggests a series of horizontal gene transfer events as the origin of their broad metabolic and ecological versatility (Wagner et al. 1998; Müller et al. 2015). Supporting this hypothesis, within the tree of life, the sulfate reduction is not only observed in deeply branched classes within *Thermodesulfobacteria* (e.g., *Thermodesulfobacterium*) but also is found in the nitrifying group *Nitrospira* (e.g., *Thermodesulfovibrio*) and spore-forming *Firmicutes* within the *Terrabacteria* group (e.g., *Desulfotomaculum*) and the best-studied group of *Deltaproteobacteria* (e.g., *Desulfovibrio*). In the Archaea domain are representatives from both classes (Euryarchaeota (e.g., *Archaeoglobus*) and Crenarchaeota (*Thermocladium* and *Caldivirga*) (Canfield et al. 2005)), suggesting again either a very deep origin or several events of horizontal gene transfer.

5. Elemental sulfur-reducing microorganisms (ESRM)

Although elemental sulfur is not a good substrate for enzymatic reactions due to its low solubility in water (5 μg/L at 47 °C) (Caspi et al. 2012), the ability to reduce elemental sulfur is widespread among several members of both prokaryotic domains of life. However, there are several biochemical and physiological differences in how sulfur is used and whether sulfur reduction serves as the principal energy acquisition process or whether it is an alternative one. Chemolithoautotrophs use a heterotrophic reduction of elemental sulfur as the primary energy source; this particular metabolism is mostly found in very deep-branched hyperthermophilic archaea from the Crenarchaeota division (TACK) (e.g., *Pyrodictium, Pyrobaculum, Sulfolobus*) including a methanogenic member (Canfield et al. 2005), e.g., *Methanopyrus, Methanobacterium*, and *Methanothermus*. It is also found in some *Deltaproteobacteria* (e.g., *Desulfurella acetivorans, Sulfospirillum arcachonense*). Although elemental sulfur is the main electron acceptor in this type of metabolism, other inorganic compounds such as nitrate, iron(III), or even oxygen can also be used. For the second group, it has been proposed that respiration of elemental sulfur constitutes an alternative metabolism for some versatile strains of *Gammaproteobacteria* such as *Wolinella, Pseudomonas*, and *Shewanella* (Canfield et al. 2005) as well as the chemolithotroph *Acidithiobacillus ferrooxidans*, within the recent proposed class *Zetaproteobacteria* (Hoshino et al. 2016). The phylogenetic position of the deep-branched chemolithoautotrophs, as well as the unspecific nature of the reaction center, suggests that very early life used this metabolic pathway, although the second group seems to go back to ancient enzymes when there were no other sources of food.

Integrating the Sulfur Cycle Within the Microbial Mat Model

As explained in detail above, each of the metabolic reactions related to the sulfur cycle (SC) is carried out by physiologically and phylogenetically diverse groups of microorganisms that live in widely varied environments and differ in their abilities to use various sulfur compounds. The great complexity of the biogeochemical sulfur cycle at the global scale is shown in Fig. 6.1, where the most important organic and inorganic S compounds derived from biogeochemical processes are arranged according to the standard Gibbs free energy of formation and the boxes summarize the sulfur metabolic guilds. The compilation of the described genera for each guild along with their corresponding complete genome sequences is found in supplementary information in De Anda et al. (2017) (http://gigadb.org/dataset/100357).

The great complexity of the sulfur cycle can be studied by considering the in microbial mats as ecological model. Microbial mats are compartmentalized organizations that have evolved over more than 3,000 Ma into the complex ecosystems that we know today (Herman and Kump 2005). Functionally, microbial mats are self-sufficient structures that support most of the major biogeochemical cycles within a vertical dimension of only a few millimeters in a multilayered space (Pinckney and Paerl 1997). The metabolic guilds of microbial mats live in close proximity, exchanging nutrients and by-products such as organic matter, sulfide, and sulfate derived from photosynthesis, sulfate reduction, and sulfide oxidation, respectively. In this way, by creating an efficient turnover of major electron acceptor/donors, sulfur gradients contribute to a fine vertical distribution that sustain communities within which the sulfur cycle is the main crossing point of the major biogeochemical cycles (Dillon et al. 2007; Herman and Kump 2005; Prieto-Barajas et al. 2017; van Gemerden 1993; Visscher and Stolz 2005). Here, we classify the metabolic guilds within microbial mats according to their method of obtaining energy and carbon for biomass, including (i) photoautotrophs, photosynthetic cyanobacteria, PSB, and GSB; (ii) chemoautotrophs, sulfur-oxidizing microorganisms such as CLSB; and finally (iii) heterotrophs, sulfate-reducing bacteria, SRB (van Gemerden 1993; Hubas et al. 2011; Bolhuis et al. 2014).

The most representative reactions performed by these sulfur metabolic guilds within a typical microbial mat are shown schematically in Fig. 6.2a. In this conceptual model, the upper layer of the mat is represented primarily by photosynthetic cyanobacteria that take advantage of abundant resources: sunlight and H_2O, which are used as source of energy and electrons (respectively) to fix CO_2. The oxygen and organic matter produced by photosynthesis rapidly diminish in deeper layers (Harris et al. 2013). The turnover of fixed carbon is caused by fermentative and heterotrophic components such as SRB that use inorganic sulfate as an external electron acceptor to oxidize organic matter compound substrates, resulting in the production of sulfide. The latter by-product, as well as other reduced sulfur compounds, is used by PSB and CLSB as electron donors for anaerobic phototrophic and chemotrophic growth (respectively). The sulfate produced in turn is used by SRB.

Biogeochemical Sulfur cycle at global scale

S-compound used as source of Carbon (C), Nitrogen (N), Energy (E), Electron donor °, unique source**	$\Delta_f G°$ of each S-compound (kcal/mol)	S-compound used as substrate for fermentation (F) or terminal electron acceptor in respiratory processes (R)
Thermithiobacillus tepidarius, Acidithiobacillus ferrooxidans, Acidithiobacillus thiooxidans (E)** *Thioalkalivibrio thiocyanoxidans, Thiobacillus aquaesulis* (E)**	$S_4O_6^{2-}$ Tetrathionate -245.5	SRB (R) Some strains of *Pseudomonas aeruginosa, Shewanella putrefaciens, Acidithiobacillus ferrooxidans* and some Enterobacteriaceae: *Salmonella, Proteus, Citrobacter, Edwardsiella* (R)
	SO_4^{2-} Sulfate -179.2	
Ruegeria pomeroyi DSS-3 (CE), *Roseovarius nubinhibens ISM* (CE), *Chromohalobacter salexigens DSM 3043* (CE)**	$C_3H_4O_6S$ Sulfolactate -176.5	
Ralstonia eutropha H16 (CE)**	$C_2H_2O_5S$ Sulfoacetate -160.6	SRB (R)
Acidianus ambivalens° *Chlorobium limicola*° *Allochromatium vinosum* and *Thiocapsa roseopersicina*° *Thiobacillus thioparus Thiobacillus denitrificans, Beggiatoa*°	SO_3^{2-} Sulfite -127.4	SRB (F)
Paracoccus pantotrophus (E)**	$C_3H_6NO_5S$ L-cysteate -119.7	
CLSB° SOM° GSB° PSB°	$S_2O_3^{2-}$ Thiosulfate -112.7	
Methylosulfonomonas, Marinosulfonomonas, and some strains of *Hyphomicrobium* and *Methylobacterium* (CE)	CH_3O_3S Methane sulfonate -71.5	*Desulfonispora thiosulfatigenes* GKNTAU (F) *Bilophila wadsworthia RZATAU* (R)
Pseudomonas aeruginosa TAU-5 (CE) *Castellaniella defegrans NKNTAU* (E**) *Paracoccus denitrificans NKNIS* (E**) *Paracoccus pantotrophus NKNCYSA* (E**)	$C_2H_7NO_3S$ Taurine -49.9	
Arthrobacter methylotrophus (CE)** *Arthrobacter sulfonivorans* (CE)** *Hyphomicrobium sulfonivorans* (CE)**	$C_2H_6O_2S$ Dimethyl sulfone -11.3	
Ralstonia eutropha H16 (N)**	$C_3H_9NO_3S$ Homotaurine -7.3	
Ruegeria pomeroyi DSS-3 (CE)	$C_4H_7O_2S$ Methylthio propanoate -3.2	SR (R) SRB/SR (R)
Thiobacillus thioparus TK-m (E) *Paracoccus denitrificans* (E)	CS_2 Carbondisulfide 0.0	
SR/SO° CLSB° SOM° GSB° PSB°	S° Elemental sulfur 8.8	Some members of the genus: *Methanopyrus, Methanobacterium, Methanothermus,* and *Methanococcus* (R)
CLSB° GSB° PSB° *Pseudanabaena limnetica*°	H_2S Sulfide 13.9	
Sulfitobacter sp. EE-36, Marinomonas sp. MED121, Marinomonas sp. MWYL1, Ruegeria pomeroyi DSS-3, Roseovarius nubinhibens ISM: (CE)	CH_4S Methanethiol 36.7	Some *Methanosarcina* species(R)
	$C_5H_{10}O_2S$ DMSP 37.5	
Thiobacillus thioparius, TK-m, Thiobacillus thioparius, T5 and *Hyphomicrobium sp* (CE)	C_2H_6S DMS 63.1	

pH 7.3 ionic strength of 0.25, 298 K

CLSB	Color-less Sulfur Bacteria: 24 genera (*i.e Beggiatoa, Thiomargarita, Thiobacillus*)
SOM	Sulfur Oxidizing Microorganims: 12 genera (*i.e Thermithiobacillus, Acidithiobacillus*)
GSB	Green Sulfur Bacteria: 9 genera (*i.e Clorobaculum, Chloroflexus, Chlorobium*)
PSB	Purple Sulfur Bacteria: 25 genera (*i.e Ectothiorhodospira, Cromatium*)
SRB	Sulfate Reducing Bacteria: 40 genera (*i.e Desulfovibrio, Desulfotomaculum, Desulfotignum*)
SR	Elemental-Sulfur Reducing microorganisms: 19 genera (*i.e Sulfospirillum, Desulfurella, Thermoproteus*)
ESR/SO	Elemental-Sulfur Reducing and Sulfur Oxidizing microorganisms: includes 4 genera: *Sulfolobus, Acidianus, Aquifex, Thermoplasma*
SRB/ESR	Sulfate Reducing Bacteria and Elemental Sulfur Reducing microorganism: 42 genera (i. e *Thermovirga, Pyrolobus*)

Fig. 6.1 Sulfur cycle at the global scale, where the most important organic and inorganic S compounds derived from biogeochemical processes are arranged according to the standard Gibbs free energy of formation. Colored boxes indicate the metabolic guilds involved in the metabolism of S compounds in oxidation (i.e., CLSB, SOM, PSB, and GSB) or reduction (SR, SRB) processes, which are summarized in Fig. 6.2. (Figure from De Anda et al. (2017))

In this context, the compartmentalization of microbial mats provides clear, natural boundaries that evoke the concept of a "minimum ecosystem." That is, specific parts of the cycle may be seen as parts of a whole. For instance, the redox level, reduced-oxidized compounds, or even genes and enzymes implicated in certain routes can be used as ecosystem boundaries. These assemblies set up a unit that represents the minimum ecosystem with minimum requirements for functionality, and therefore this can be applied as an ecological model to understand the intersection of the biogeochemical cycles (Guerrero et al. 2002). Considering the difficulties in delimiting the parameters that define a whole biogeochemical cycle, including its limits and scope or what elements should be considered as necessary to each cycle, we focus on metabolic coupling within microbial mats to highlight the cycles interplay at taxonomic and metabolic levels (Fig. 6.2a). In this schematic representation, the colored circles represent the metabolic guilds within the microbial mats ranging from the outer circle (layer 1), represented by cyanobacteria, followed by CLSB (layer2), PSB (layer 3), GSB (layer 4), and finally SRB (layer 5). The colors within each slice represent the metabolic intersection of the biogeochemical cycles (C, H, O, N, S, P, Fe). Below we explain how this interplay is represented, moving from the outer to the inner cycle. In layer 1, *Cyanobacteria* carry out CO_2 fixation by Calvin cycle (C), releasing oxygen (O) using water H_2O as electron donor (H). Some *Cyanobacteria* can perform nitrogen fixation (N), i.e., *Oscillatoria limnetica* that uses H_2S as electron donor. Sulfur's intersection with the P cycle is seen in the biosynthesis of the sulfolipid sulfoquinovosyl diacylglycerol (SQDG) by some *Cyanobacteria* (*Synechococcus* and *Synechocystis*) and *Bacillus coahuilensis* at CCB (Alcaraz et al. 2008). The CLSB (layer 2 within the mat model) require an inorganic source of energy and obtain their cell carbon from the fixation of carbon dioxide by Calvin cycle (C). Using sulfide both as energy source and electron donor with oxygen (O), nitrate, or nitrite as terminal acceptors, the nitrogen oxides are reduced to nitrogen (N, denitrification) to obtain energy for growth. For example, *Thiobacillus denitrificans* is able to grow with O, nitrate, or nitrite as terminal electron donors. *Thiobacillus ferrooxidans* reduces ferric iron (Fe) under anaerobic conditions. The PSB (layer 3) use intermediate products of organic matter degradation by cyanobacteria and inorganic sulfur compounds produced by sulfate-reducing bacteria as sulfide or molecular H_2/H as electron donors for the light-dependent reduction of CO_2 to cell material (C). Synthesis of SQDG is also reported in PSB such as in *Rhodobacter sphaeroides*. In contrast to the PSB, the GSB (layer 4) use the reverse TCA cycle to fix CO_2; however, all the biogeochemical intersections are shared as above. Finally, as mentioned above, dissimilatory sulfate reduction by SRB (layer 5) is an energy-gaining process in which the reduction of sulfate to sulfide is linked to either oxidation of an electron donor organic compound (C) or hydrogen (H). Pyrite (FeS_2) and iron monosulfide (FeS) are formed as the result of organic matter degradation by sulfate-reducing bacteria. In addition, there are also strains of SRB capable of fixing N using the nitrogenase (*nif*) complex. This entire complexity has been found in each of the sequenced microbial mats and stromatolites at CCB (De Anda 2017, submitted; Desnues et al. 2008; Nitti et al. 2012; Peimbert et al. 2012)

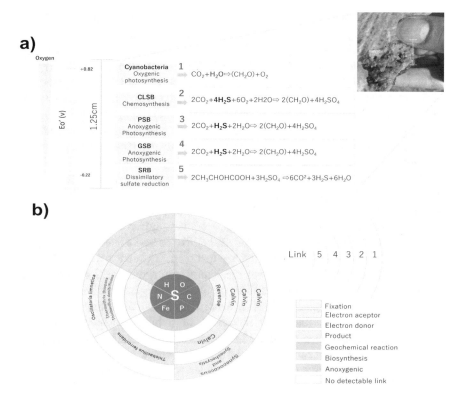

Fig. 6.2 Microbial mat model. (**a**) Representation of the vertical stratification of microorganisms where physiological, metabolic, and geochemical interactions according to redox potential are illustrated. (**b**) Diagram of the intersection of the biogeochemical cycles in a microbial mat, where the sulfur cycle is the guiding axis of the intersection. The numbers indicate microbial mat guilds from top to bottom, 1–5 indicating *Cyanobacteria*, colorless sulfur bacteria, purple sulfur bacteria, green sulfur bacteria, and finally sulfate-reducing bacteria, respectively

Capturing the Importance of Global Sulfur Cycle with a Single Value

The phylogenetic and metabolic complexity of sulfur-related microorganisms that extends over the tree of life has hampered their detection using a single gene-targeted marker, especially considering the great breadth of the metabolic pathways involved. Therefore, over the past decades, the use of several marker genes has been crucial to advancing the understanding of evolutionary ecology of sulfur metabolism in environmental samples (Dar et al. 2007; Frigaard and Bryant 2004; Hügler et al. 2010; Loy et al. 2009; Meyer and Kuever 2007; Vladár et al. 2008; Watanabe et al. 2013). In the current omics era, emerging data have confirmed the importance and widespread distribution of organic sulfur assimilation in the ocean (Delmont et al. 2011; Todd et al. 2011), sulfur redox metabolism in hot springs (Jiménez et al. 2012; Tang et al. 2013), stromatolites and microbial mats (Breitbart et al. 2009; Khodadad and Foster 2012; Warden et al. 2016), biofilms (Wilbanks et al. 2014), and hydrothermal vents (Nakai et al. 2011). In addition, similar genomics-based studies have highlighted the genetic and functional bases for the recycling of sulfur in oxygen minimum zones (Canfield et al. 2010) and have provided insights into the genetic repertoire of poorly characterized microorganisms that reside deep within Earth's crust and sediments (Mason et al. 2014; Jungbluth et al. 2017). However, despite the great advances in microbial ecology and the explosion of high-throughput sequencing data, our ability to understand and integrate the global biogeochemical cycles is still limited. Recently, we have developed a computational algorithm aimed at classifying, evaluating, and comparing large-scale "omics" samples according to their metabolic machinery. To test our algorithm, we performed a reconstruction of the biogeochemical sulfur cycle, resulting in a comprehensive, manually curated inventory of the biotic players involved. This inventory includes genes, molecular pathways, enzymes, sulfur compounds, and the microorganisms involved (De Anda et al. 2017). To the best of our knowledge, this is the first attempt to use not only individual marker genes but also a comprehensive analysis of the entire sulfur-processing machinery to integrate the biotic and abiotic processes involved in the mobilization of inorganic-organic sulfur compounds through microbial-catalyzed reactions at a global scale. Our algorithm was tested at a multi-genomic scale (including thousands of public completely sequenced genomes and metagenomes). The results strongly highlight the broad ability of our proposed algorithm to accurately detect the enrichment of sulfur metabolism even in genomes derived from environmental samples without any cultured representatives (a.k.a. "microbial dark matter"). In this way, we were able to detect the most important organisms and environments worldwide in which the overall sulfur machinery was overrepresented. Consistent with oldest geological record for the emergence of life on Earth, we were able to identify microbial mats from Cuatro Cienegas (see Chap. 8), hydrothermal vents, and hot springs as the current environments that hold the complete repertoire of sulfur metabolic pathways (including the mobilization of organic and inorganic sulfur compounds) in worldwide environments (De Anda et al., 2017).

Conclusions and Remarks

The ongoing debates over when life emerged on Earth or which signal is the oldest evidence of life have been subjected to continue renewal of evidence that had placed the origin of life earlier and earlier within geological time scales. However, despite the fragmentary nature of the fossil record, newly obtained geologic evidence indicates that hot springs, microbial mats, and hydrothermal vents were the primordial environments with the required conditions for the beginning of the life based on the sulfur cycle. Sulfur continues to play a critical role in modern analogues, such as the microbial mats from Cuatro Cienegas (see Chap. 8). In agreement with the geological record, molecular and genomic data, along with computational in silico modeling, also point out the critical role of sulfur in the beginning of life, suggesting that the use of sulfur compounds was probably the most ancient metabolism that appeared on Earth, a process that continues to play a critical role in the modern day.

We also recapitulate the great metabolic diversity of the sulfur cycle within the recent tree of life, and we use microbial mats as ecological models of early biogeochemical evolution; since they are nearly closed and self-sustaining ecosystems that encompass the major biogeochemical cycles, trophic levels, and food webs in a vertically laminated pattern. Our conceptualized microbial mat model allows us to highlight the important metabolic links of the major biogeochemical cycles performed by five sulfur metabolic guilds that are widespread in nature and encountered at the millimeter scale within microbial mats. Our emerging comparative framework and emphasis on historical patterns are helping to bridge barriers among organism-based research, community studies, phylogenomic analysis, mathematical modeling, and paleogeochemical data, pointing out the critical role of sulfur in the early history of life.

Acknowledgments This work constitutes a partial fulfillment requirement for the Ph.D. degree of Valerie De Anda at the graduate program Doctorado en Ciencias Biomédicas of the Universidad Nacional Autónoma de México who received fellowship 356 832 from Consejo Nacional de Ciencia y Tecnología (CONACYT). The authors acknowledge the founding of WWF-Alianza Carlos Slim, as well as support by a Sep Conacyt Project to VS and LEE 1101OL34. (aqui hablar sobre que es proyecto del azufre). The manuscript was written during a sabbatical leave of LEE and VSS in the University of Minnesota in Peter Tiffin and Michael Travisano laboratories, with support of the program PASPA- DGAPA, UNAM.

We would like to acknowledge Peter Stadler whose valuable comments greatly improve the manuscript.

References

Alcaraz LD, Olmedo G, Bonilla G et al (2008) The genome of Bacillus coahuilensis reveals adaptations essential for survival in the relic of an ancient marine environment. Proc Natl Acad Sci U S A 105:5803–5808. https://doi.org/10.1073/pnas.0800981105

Allwood AC, Walter MR, Kamber BS et al (2006) Stromatolite reef from the Early Archaean era of Australia. Nature 441:714–718. https://doi.org/10.1038/nature04764

Aoyama S, Ueno Y (2018) Multiple sulfur isotope constraints on microbial sulfate reduction below an Archean seafloor hydrothermal system. Geobiology 16:107–120. https://doi.org/10.1111/gbi.12268

Bolhuis H, Cretoiu MS, Stal LJ (2014) Molecular ecology of microbial mats. FEMS Microbiol Ecol 90:335–350. https://doi.org/10.1111/1574-6941.12408

Breitbart M, Hoare A, Nitti A et al (2009) Metagenomic and stable isotopic analyses of modern freshwater microbialites in Cuatro Ciénegas, Mexico. Environ Microbiol 11:16–34

Canfield D, Kristensen E, Bo T (2005) The sulfur cycle, 1st edn. Aquat. Geomicrobiol. (Advances Mar. Biol) Elsevier Academic Press

Canfield D, Stewart F, Thamdrup B et al (2010) A cryptic sulfur cycle in oxygen-minimum-zone waters off the Chilean coast. Science 330:1375–1378. https://doi.org/10.1126/science.1196889

Caspi R, Altman T, Dreher K et al (2012) The MetaCyc database of metabolic pathways and enzymes and the BioCyc collection of pathway/genome databases. Nucleic Acids Res 40:D742–D753. https://doi.org/10.1093/nar/gkr1014

Dai J (2017) New insights into a hot environment for early life. Environ Microbiol Rep 9:1–26

Dalton R (2004) Fresh study questions oldest traces of life in Akilia rock. Nature 429:688

Dar SA, Yao L, van Dongen U et al (2007) Analysis of diversity and activity of sulfate-reducing bacterial communities in sulfidogenic bioreactors using 16S rRNA and dsrB genes as molecular markers. Appl Environ Microbiol 73:594–604. https://doi.org/10.1128/AEM.01875-06

De Anda V, Zapata-Peñasco I, Poot-Hernandez AC et al (2017) MEBS, a software platform to evaluate large (meta)genomic collections according to their metabolic machinery: unraveling the sulfur cycle. Gigascience 6:1–17. https://doi.org/10.1093/gigascience/gix096

Delmont TO, Malandain C, Prestat E et al (2011) Metagenomic mining for microbiologists. ISME J 5:1837–1843. https://doi.org/10.1038/ismej.2011.61

Desnues C, Rodriguez-Brito B, Rayhawk S et al (2008) Biodiversity and biogeography of phages in modern stromatolites and thrombolites. Nature 452:340–343. https://doi.org/10.1038/nature06735

Dillon JG, Fishbain S, Miller SR et al (2007) High rates of sulfate reduction in a low-sulfate hot spring microbial mat are driven by a low level of diversity of sulfate-respiring microorganisms. Appl Environ Microbiol 73:5218–5226. https://doi.org/10.1128/AEM.00357-07

Djokic T, VanKranendonk MJ, Campbel KA et al (2017) Earliest signs of life on land preserved in ca. 3.5 Ga hot spring deposits. Nat Commun 8:1–8. https://doi.org/10.1038/ncomms15263

Dodd MS, Papineau D, Grenne T et al (2017) Evidence for early life in Earth's oldest hydrothermal vent precipitates. Nature 543:60–64. https://doi.org/10.1038/nature21377

Friedrich CG, Rother D, Bardischewsky F et al (2001) Oxidation of reduced inorganic sulfur compounds by bacteria: emergence of a common mechanism? Appl Environ Microbiol 67:2873–2882. https://doi.org/10.1128/AEM.67.7.2873-2882.2001

Frigaard N-U, Bryant DA (2004) Seeing green bacteria in a new light: genomics-enabled studies of the photosynthetic apparatus in green sulfur bacteria and filamentous anoxygenic phototrophic bacteria. Arch Microbiol 182:265–276. https://doi.org/10.1007/s00203-004-0718-9

Frigaard N, Bryant DA (2008) Genomic and evolutionary perspectives on sulfur metabolism in green sulfur bacteria. In: Dahl C, Cornelius F (eds) Microbial sulfur metabolism. Springer, Berlin, Heidelberg, pp 60–76

Garcia AK, Schopf JW, Yokobori S et al (2017) Reconstructed ancestral enzymes suggest long-term cooling of Earth's photic zone since the Archean. Proc Natl Acad Sci 114:4619–4624. https://doi.org/10.1073/pnas.1702729114

Gaucher EA, Kratzer JT, Randall RN (2010) Deep phylogeny--how a tree can help characterize early life on earth. Cold Spring Harb Perspect Biol 2:1–16. https://doi.org/10.1101/cshperspect.a002238

Ghosh W, Dam B (2009) Biochemistry and molecular biology of lithotrophic sulfur oxidation by taxonomically and ecologically diverse bacteria and archaea. FEMS Microbiol Rev 33:999–1043. https://doi.org/10.1111/j.1574-6976.2009.00187.x

Goldford JE, Hartman H, Smith TF, Segrè D (2017) Remnants of an ancient metabolism without phosphate. Cell 168:1126–1134.e9. https://doi.org/10.1016/j.cell.2017.02.001

Grassineau NV, Nisbet EG, Bickle MJ et al (2001) Antiquity of the biological sulphur cycle: evidence from sulphur and carbon isotopes in 2700 million-year-old rocks of the Belingwe Belt, Zimbabwe. Proc R Soc B Biol Sci 268:113–119. https://doi.org/10.1098/rspb.2000.1338

Guerrero R, Piqueras M, Berlanga M (2002) Microbial mats and the search for minimal ecosystems. Int Microbiol 5(4):177–188. Epub 2002 Nov 7. Review. PubMed PMID: 12497183. https://link.springer.com/article/10.1007%2Fs10123-002-0094-8

Harris JK, Caporaso JG, Walker JJ et al (2013) Phylogenetic stratigraphy in the Guerrero Negro hypersaline microbial mat. ISME J 7:50–60. https://doi.org/10.1038/ismej.2012.79

Herman EK, Kump LR (2005) Biogeochemistry of microbial mats under Precambrian environmental conditions: a modelling study. Geobiology 3:77–92

Hoshino T, Kuratomi T, Morono Y et al (2016) Ecophysiology of Zetaproteobacteria associated with shallow hydrothermal iron-oxyhydroxide deposits in Nagahama Bay of Satsuma Iwo-Jima, Japan. Front Microbiol 6:1554. https://doi.org/10.3389/fmicb.2015.01554

Hubas C, Jesus B, Passarelli C, Jeanthon C (2011) Tools providing new insight into coastal anoxygenic purple bacterial mats: review and perspectives. Res Microbiol 162:858–868

Hug LA, Baker BJ, Anantharaman K et al (2016) A new view of the tree of life. Nat Microbiol 1:16048. https://doi.org/10.1038/nmicrobiol.2016.48

Hügler M, Gärtner A, Imhoff JF (2010) Functional genes as markers for sulfur cycling and CO_2 fixation in microbial communities of hydrothermal vents of the Logatchev field. FEMS Microbiol Ecol 73:526–537. https://doi.org/10.1111/j.1574-6941.2010.00919.x

Jiménez DJ, Andreote FD, Chaves D et al (2012) Structural and functional insights from the metagenome of an acidic hot spring microbial planktonic community in the Colombian Andes. PLoS One 7:e52069. https://doi.org/10.1371/journal.pone.0052069

Jungbluth SP, Glavina del Rio T, Tringe SG et al (2017) Genomic comparisons of a bacterial lineage that inhabits both marine and terrestrial deep subsurface systems. PeerJ 5:e3134. https://doi.org/10.7717/peerj.3134

Khodadad CLM, Foster JS (2012) Metagenomic and metabolic profiling of nonlithifying and lithifying stromatolitic mats of Highborne Cay, The Bahamas. PLoS One 7:e38229. https://doi.org/10.1371/journal.pone.0038229

Koonin EV, Martin W (2005) On the origin of genomes and cells within inorganic compartments. Trends Genet 21:647–654. https://doi.org/10.1016/j.tig.2005.09.006

Lake JA, Skophammer RG, Herbold CW, Servin JA (2009) Genome beginnings: rooting the tree of life. Philos Trans R Soc Lond B Biol Sci 364:2177–2185. https://doi.org/10.1098/rstb.2009.0035

Lau MC, Kieft TL, Kuloyo O et al (2016) An oligotrophic deep-subsurface community dependent on syntrophy is dominated by sulfur-driven autotrophic denitrifiers. Proc Natl Acad Sci U S A 113:E7927–E7936

Loy A, Duller S, Baranyi C et al (2009) Reverse dissimilatory sulfite reductase as phylogenetic marker for a subgroup of sulfur-oxidizing prokaryotes. Environ Microbiol 11:289–299. https://doi.org/10.1111/j.1462-2920.2008.01760.x

Martin WF, Thauer RK (2017) Energy in ancient metabolism. Cell 168:953–955. https://doi.org/10.1016/j.cell.2017.02.032

Mason OU, Scott NM, Gonzalez A et al (2014) Metagenomics reveals sediment microbial community response to Deepwater Horizon oil spill. ISME J 8:1464–1475. https://doi.org/10.1038/ismej.2013.254

Meyer B, Kuever J (2007) Molecular analysis of the distribution and phylogeny of dissimilatory adenosine-5′-phosphosulfate reductase-encoding genes (aprBA) among sulfur-oxidizing prokaryotes. Microbiology 153:3478–3498. https://doi.org/10.1099/mic.0.2007/008250-0

Müller AL, Kjeldsen KU, Rattei T et al (2015) Phylogenetic and environmental diversity of DsrAB-type dissimilatory (bi)sulfite reductases. ISME J 9:1152–1165. https://doi.org/10.1038/ismej.2014.208

Muyzer G, Stams AJM (2008) The ecology and biotechnology of sulphate-reducing bacteria. Nat Rev Microbiol 6:441–454

Nakai R, Abe T, Takeyama H, Naganuma T (2011) Metagenomic analysis of 0.2-μm-passable microorganisms in deep-sea hydrothermal fluid. Mar Biotechnol 13:900–908. https://doi.org/10.1007/s10126-010-9351-6

Nitti A, Daniels CA, Siefert J et al (2012) Spatially resolved genomic, stable isotopic, and lipid analyses of a modern freshwater microbialite from Cuatro Ciénegas, Mexico. Astrobiology 12:685–698. https://doi.org/10.1089/ast.2011.0812

Noffke N, Christian D, Wacey D, Hazen RM (2013) Microbially induced sedimentary structures recording an ancient ecosystem in the *ca.* 3.48 billion-year-old dresser formation, Pilbara, Western Australia. Astrobiology 13:1103–1124. https://doi.org/10.1089/ast.2013.1030

Nutman AP, Bennett VC, Friend CRL et al (2016) Rapid emergence of life shown by discovery of 3,700-million-year-old microbial structures. Nature 537:535–538. https://doi.org/10.1038/nature19355

Olson KR, Straub KD, Straub KD (2016) The role of hydrogen sulfide in evolution and the evolution of hydrogen sulfide in metabolism and signaling. Physiology 31:60–72. https://doi.org/10.1152/physiol.00024.2015

Ozuolmez D, Na H, Lever M et al (2015) Methanogenic archaea and sulfate reducing bacteria co-cultured on acetate: teamwork or coexistence? Front Microbiol 6:492. https://doi.org/10.3389/fmicb.2015.00492

Peimbert M, Alcaraz LD, Bonilla-Rosso G et al (2012) Comparative metagenomics of two microbial mats at cuatro ciénegas basin I: ancient lessons on how to cope with an environment under severe nutrient stress. Astrobiology 12:648–658. https://doi.org/10.1089/ast.2011.0694

Pereira IA, Ramos AR, Grein F et al (2011) A comparative genomic analysis of energy metabolism in sulfate reducing bacteria and archaea. Front Microbiol 2:69. https://doi.org/10.3389/fmicb.2011.00069

Philippot P, Van Zuilen M, Lepot K et al (2007) Early archaean microorganisms preferred elemental sulfur, not sulfate. Science 317:1534–1537. https://doi.org/10.1126/science.1145861

Pinckney JL, Paerl HW (1997) Anoxygenic photosynthesis and nitrogen fixation by a microbial mat community in a bahamian hypersaline lagoon. Appl Environ Microbiol 63:420–426

Plugge C, Zhang W, Scholten J, Stams A (2011) Metabolic flexibility of sulfate-reducing bacteria. Front Microbiol 2:81. https://doi.org/10.3389/fmicb.2011.00081

Prieto-Barajas CM, Valencia-Cantero E, Santoyo G (2017) Microbial mat ecosystems: structure types, functional diversity, and biotechnological application. Electron J Biotechnol 31:48–56. https://doi.org/10.1016/j.ejbt.2017.11.001

Robertson LA, Kuenen JG (2006) The colorless sulfur bacteria. In: Dworkin M, Falkow S, Rosenberg E et al (eds) The prokaryotes: volume 2: ecophysiology and biochemistry. Springer, New York, NY, pp 985–1011

Schopf JW (1993) Microfossils of the Early Archean Apex chert: new evidence of the antiquity of life. Science 260:640–646. https://doi.org/10.1126/science.260.5108.640

Schopf JW, Kitajima K, Spicuzza MJ et al (2018) SIMS analyses of the oldest known assemblage of microfossils document their taxon-correlated carbon isotope compositions. Proc Natl Acad Sci 115:53–58. https://doi.org/10.1073/pnas.1718063115

Semenov SN, Kraft LJ, Ainla A et al (2016) Autocatalytic, bistable, oscillatory networks of biologically relevant organic reactions. Nature 537:656–660. https://doi.org/10.1038/nature19776

Shen Y, Buick R, Canfield D (2001) Isotopic evidence for microbial sulphate reduction in the early Archaea era. Nature 410:77–81

Shen Y, Farquhar J, Masterson A et al (2009) Evaluating the role of microbial sulfate reduction in the early Archean using quadruple isotope systematics. Earth Planet Sci Lett 279:383–391. https://doi.org/10.1016/j.epsl.2009.01.018

Staley JT (2002) The metabolism of Earth's first organisms. In: American Astronomical Society meeting abstracts, p 1221

Tang K, Liu K, Jiao N et al (2013) Functional metagenomic investigations of microbial communities in a shallow-sea hydrothermal system. PLoS One 8:e72958. https://doi.org/10.1371/journal.pone.0072958

Tashiro T, Ishida A, Hori M et al (2017) Early trace of life from 3.95 Ga sedimentary rocks in Labrador, Canada. Nature 549:516–518. https://doi.org/10.1038/nature24019

Tice MM, Lowe DR (2004) Photosynthetic microbial mats in the 3,416-Myr-old ocean. Nature 431:549–552. https://doi.org/10.1038/nature02920.1

Todd JD, Curson ARJ, Kirkwood M et al (2011) DddQ, a novel, cupin-containing, dimethyl-sulfoniopropionate lyase in marine roseobacters and in uncultured marine bacteria. Environ Microbiol 13:427–438. https://doi.org/10.1111/j.1462-2920.2010.02348.x

Trudinger PA (1992) Bacterial sulfate reduction: current status and possible origin. In: Schidlowski M et al (eds) Early organic evolution: implications for mineral and energy resources. Springer-Verlag, Berlin, Heidelberg, pp 367–377

Ueno Y, Ono S, Rumble D, Maruyama S (2008) Quadruple sulfur isotope analysis of ca. 3.5 Ga Dresser Formation: new evidence for microbial sulfate reduction in the early Archean. Geochim Cosmochim Acta 72:5675–5691. https://doi.org/10.1016/j.gca.2008.08.026

van Gemerden H (1993) Microbial mats: a joint venture. Mar Geol 113:3–25. https://doi.org/10.1016/0025-3227(93)90146-M

Van Kranendonk MJ, Philippot P, Lepot K et al (2008) Geological setting of Earth's oldest fossils in the ca. 3.5 Ga Dresser Formation, Pilbara Craton, Western Australia. Precambrian Res 167:93–124. https://doi.org/10.1016/j.precamres.2008.07.003

Van Kranendonk MJ, Deamer DW, Djokic T (2017) Life springs. Sci Am 317:28–35. https://doi.org/10.1038/scientificamerican0817-28

Visscher PT, Stolz JF (2005) Microbial mats as bioreactors: populations, processes, and products. Palaeogeogr Palaeoclimatol Palaeoecol 219:87–100. https://doi.org/10.1016/j.palaeo.2004.10.016

Vladár P, Rusznyák A, Márialigeti K, Borsodi AK (2008) Diversity of sulfate-reducing bacteria inhabiting the rhizosphere of Phragmites australis in Lake Velence (Hungary) revealed by a combined cultivation-based and molecular approach. Microb Ecol 56:64–75. https://doi.org/10.1007/s00248-007-9324-0

Wächtershäuser G (1990) The case for the chemoautotrophic origin of life in an iron-sulfur world. Orig Life Evol Biosph 20:173–176. https://doi.org/10.1007/BF01808279

Wächtershäuser G (2008) Iron-sulfur world. In: Begley T (ed) Wiley encyclopedia of chemical biology. American Cancer Society. Wiley, Chichester, pp 1–8

Wagner M, Roger AJ, Flax JL et al (1998) Phylogeny of dissimilatory sulfite reductases supports an early origin of sulfate respiration. J Bacteriol 180:2975–2982

Walter MR, Buick R, Dunlop JSR (1980) Stromatolites 3,400-3,500 Myr old from the North Pole area, Western Australia. Nature 284:443–445

Warden JG, Casaburi G, Omelon CR et al (2016) Characterization of microbial mat microbiomes in the modern thrombolite ecosystem of Lake Clifton, Western Australia using shotgun metagenomics. Front Microbiol 11:1–14

Watanabe T, Kojima H, Takano Y, Fukui M (2013) Diversity of sulfur-cycle prokaryotes in freshwater lake sediments investigated using aprA as the functional marker gene. Syst Appl Microbiol 36:436–443

Westall F, Vries ST, Nijman W et al (2006) The 3.466 Ga "Kitty's Gap Cheil," an early Archean microbial ecosystem. GSA Spec Pap 405:105–131

Wilbanks EG, Jaekel U, Salman V et al (2014) Microscale sulfur cycling in the phototrophic pink berry consortia of the Sippewissett Salt Marsh. Environ Microbiol 16:3398–3415. https://doi.org/10.1111/1462-2920.12388

Zhelezinskaia I, Kaufman AJ, Farquhar J, Cliff J (2014) Large sulfur isotope fractionations associated with Neoarchean microbial sulfate reduction. Science 346:742–744. https://doi.org/10.1126/science.1256211

Chapter 7
Toward a Comprehensive Understanding of Environmental Perturbations in Microbial Mats from the Cuatro Cienegas Basin by Network Inference

Valerie De Anda, Icoquih Zapata-Peñasco, Luis E. Eguiarte, and Valeria Souza

Contents

Abstract The Cuatro Cienegas Basin (CCB) encompasses hundreds of aquatic systems that harbor diverse microbialites with different community structure composition and with the highest level of endemism in North America. Thus, CCB represents a desert oasis of high biodiversity. Despite the great importance of this unique site, increasing demand on water for agricultural development (forage and feed livestock) was first manifested with extraction of groundwater in 2011, starting the drying process of aquifers and desertification of the Churince Lagoon. As consequence, water levels have been drastically fluctuating, affecting all ecosystem functions. This chapter reviews a recent network-based approach used to understand how the anthropogenic disturbances affect one of the most resistant microbial communities since the Archean, microbial mats.

V. De Anda · L. E. Eguiarte · V. Souza (✉)
Departamento de Ecología Evolutiva, Instituto de Ecología, Universidad Nacional Autónoma de México, Coyoacan, Mexico
e-mail: souza@unam.mx

I. Zapata-Peñasco
Instituto Mexicano del Petróleo, Mexico City, Mexico

© Springer International Publishing AG, part of Springer Nature 2018
F. García-Oliva et al. (eds.), *Ecosystem Ecology and Geochemistry of Cuatro Cienegas*, Cuatro Ciénegas Basin: An Endangered Hyperdiverse Oasis,
https://doi.org/10.1007/978-3-319-95855-2_7

Keywords Network-Inference · Aquatic systems · Biodiversity · Water Overexploitation
· Microbial mats

General Overview of CCB Hydrogeology and Its Aquifer Overexploitation

The Cuatro Cienegas Basin (CCB) is an extremely low-P aquatic environment in a naturally isolated valley in the Chihuahuan Desert (Coahuila, Mexico). CCB is listed as a Wetland of International Importance under the International RAMSAR Convention, an area of protection of flora and fauna (ANP) under Mexican government, and a priority site for Conservation of Nature by the World Wide Fund for Nature and UNESCO (Souza et al. 2012). CCB has complex hydrogeological characteristics that have been shaped by the basin's tectonic activity. Mesozoic faults associated with the opening of the Gulf of Mexico created conditions that permitted the deposition of Cretaceous carbonate rocks that now form a regional carbonate aquifer within CCB (Wolaver et al. 2012). According to isotopic data, this aquifer is affected by the presence of older deeply penetrating faults that provide endogenic fluids from Earth's mantle and lower crust to surface water, producing the unique physicochemical characteristics of the current aquatic environments (Wolaver and Diehl 2010; Wolaver et al. 2012).

The aquatic systems within CCB exhibit two distinct classes of water that reflect the origin of that water (see the hydrogeological conceptual model described in detail in Wolaver et al. 2012). The first class includes $CaSO_4$-rich waters derived from a regional carbonate aquifer discharge mixed with contributions from deeply sourced fluid from the mantle that ascend along basement-involved faults (e.g., La Becerra and Churince). The second class involves $CaHCO_3$ waters with contributions made via the carbonate aquifer mixed with locally recharged mountain precipitation (e.g., Santa Tecla).

In recent decades, reduced deepwater inputs associated with intensively irrigated crops grown for forage and livestock feed prompted the desiccation of one of the principal aquatic systems within CCB, the Churince drainage. This anthropogenic disturbance has led to dramatic shifts in habitat conditions of the system, affecting turtles, fishes (Carson et al. 2015; Hernández et al. 2017), and microbial mats (De Anda under review).

The Churince System: A Scenario for Our Study

The Churince system is unique within the CCB because it is surrounded by pure gypsum (calcium sulfate) dunes that date to the Jurassic period. It contains no calcium carbonate deposits, indicating that this marine sediment was never buried by recent sediments (Wolaver et al. 2012) or organic matter from forest litter (Minckley and Jackson 2007). This is particularly relevant in the Churince system since, as mentioned above, the water derived from the aquifer is shaped by deepwater fluids influenced by mantle geochemistry (Wolaver and Diehl 2010; Wolaver et al. 2012). Previous to 2006, the Churince system was represented by a spring and a river system that flowed freely downstream, an intermediate lagoon with constant water supply, and finally a large desiccation lagoon (Laguna Grande), the depth of which varied with evaporation (Cerritos et al. 2011).

However, in summer 2007, Laguna Grande disappeared, and the Laguna Intermedia started to have strong seasonal water fluctuations with extensive desiccation during summer and recovery in winter due to reduced evaporation (Souza et al. 2007). Laguna Intermedia has experienced extensive water depletion during hot months since summer 2011. As observed in Fig. 7.1, Lagunita Pond (lateral to Laguna Intermedia) (see Chap. 4) started to display a greater desiccation during autumn from 2014 to 2017, and during 2016 and 2017 for the first time in our records, we observed a long period of complete desiccation in which water levels recover in spring and summer.

In spring 2013 the basin's main canal at La Becerra was closed due to a sustained conservation effort to recover the wetland. This closure made Churince's near-miraculous recovery possible; we record that recovery in this study. However, after summer 2014, that recovery stopped due to the reopening of the La Becerra canal for 2 weeks, changing the recharge dynamics of the Churince wetland. During 2016–2017, the Churince only had water in the beginning of spring, and even that supply was limited, highlighting the urgency of closing all the canals that export water out of the basin as well as changing agricultural practices within the area. This water depletion in the Churince system is "a canary in the mine" that indicates larger threats due to hydrological disruption because this site is few meters higher (745 masl) than the rest of the basin (average 730 masl) due to the tilt caused by the San Marcos Sierras (Wolaver et al. 2012).

An alternative explanation for the desiccation of the Churince system is extended drought associated with global warming trends. However, we believe that this is unlikely because the rainfall at CCB has actually been, on average, somewhat higher in the last 40 years (Montiel-González et al. 2017). Thus, the desiccation of Churince is most likely associated with continuous overexploitation of the aquifer for intensive agriculture in which spring waters are diverted toward a series of canals that irrigate farms even many kilometers away of the basin. This trend started in the 1960s with the opening of deep canals in the main springs to export water to irrigated agricultural fields in the region as well as within the valley (Minckley 1969) in order to water alfalfa for cattle feed. The problem started to become even more pressing during the last 15 years, with expansion of the agricultural frontier in neighboring valleys in 2002 and the opening of many deep wells within and outside the basin (Google Earth visual assessment and CNA (Comisión Nacional del Agua) data). This acceleration of water extraction precipitated the loss of Laguna Grande and a large proportion of the basin's wetlands. This collapse occurs because of subterranean connections between the three valleys undergoing agricultural expansion,

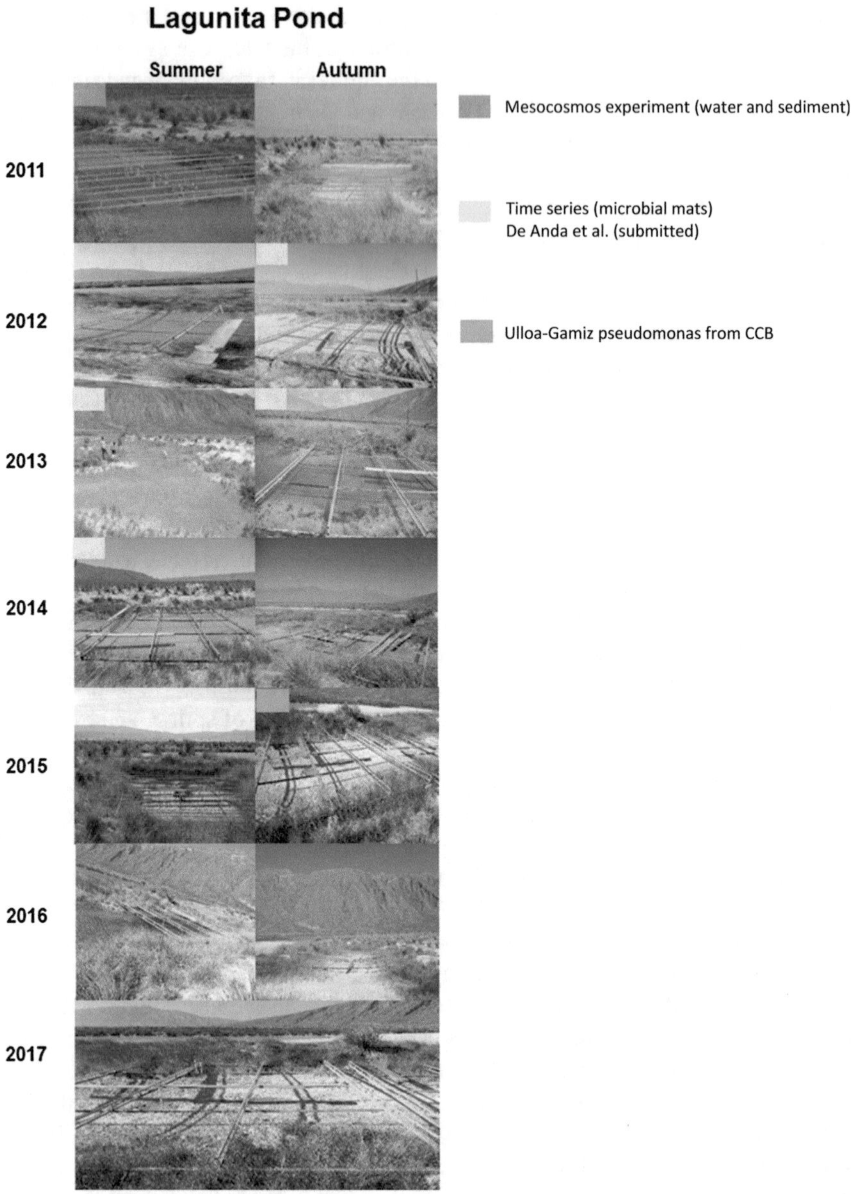

Lagunita Pond

Fig. 7.1 Water changes between autumn and spring from 2011 to 2017 within Lagunita Pond (26.84810° N, −102.14160° W), lateral to the main Churince system

including Ocampo to the north and El Hundido to the south (Fig. 7.2) (Souza et al. 2006). To our surprise, the evidence gathered by our team prompted two presidential decrees to declare two water use restrictions (vedas), El Hundido veda in 2007 and Ocampo-Cuatro Cienegas in 2013 (https://www.gob.mx/inecc/acciones-y-pro-

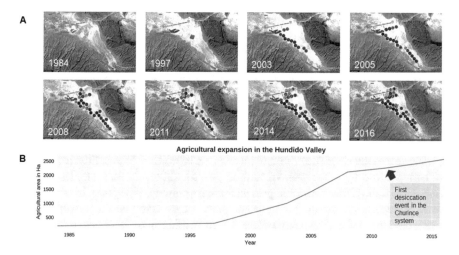

Fig. 7.2 Agricultural expansion in the Hundido Valley (26.594838° N, −102.214423° W). (**a**) Images obtained with Google Earth Image Landsat-Copernicus from 1984 to 2016. (**b**) Agricultural area in hectares was computed by calculating the surface of crop land using Google Earth

gramas/analisis-de-la-variacion-del-nivel-de-los-principales-cuerpos-de-agua-de-cuatrocienegas). The trend of the agricultural expansion in the southwestern region within CCB (the Hundido) from 1984 to 2016 can be seen in Fig. 7.2.

During our first sampling (see Winter 2012 in Fig. 7.1), we observed strong depletion of the aquifer, followed, as mentioned above, by an apparent recovery of the ecosystem in summer 2013 due to the closing of the "La Becerra" canal and the rebirth of the "El Garabatal" wetland that had been dry since the 1960s (Minckley 1969). The restoration of the water cycle in the wetland allowed Churince to recover its water flow for a small period that we were able to document (see yellow rectangles in Fig. 7.1).

Microbial Mats: A Biomarker to Understand Environmental Perturbation

Because microbial communities are the heart of every ecosystem in our planet and given the intense overexploitation of water resources within CCB, one of our main aims has been to understand how microbial systems respond to ongoing environmental disturbance so that we can establish motivation for conservation and protection of the aquifer. Considering that natural microbial communities are in constant flux even in the absence of environmental perturbations, we hypothesized that it is easier to identify the underlying mechanisms that arise from an environmental disturbance in very stable and resilient natural communities (for further information see Konopka et al. 2015).

Microbial mats are one of the oldest microbial ecosystems, having been established in the fossil record since the Archean. Thus, they are well known to be successful ecological communities that survived millions of years of environmental disruption due to their internal capacity to carry out all fundamental biogeochemical and energy transformations (van Gemerden 1993; Bolhuis et al. 2014; Prieto-Barajas et al. 2017). Therefore, they are excellent models for understanding the role of environmental disturbances on microbial communities. Our recent study (De Anda under review) evaluated microbial mats from Churince Lagoon in CCB over the course of 2 years during and after a desiccation event (see Fig. 7.1 yellow rectangles). We hypothesized that if microbial mats are indeed stable over time (regardless of perturbations by water overexploitation), we should expect that samples taken at fixed time points in space would present similar community patterns. In contrast, if there is indeed a perturbation to the community, we expected systematic differences that would then reflect the degree of resistance, resilience, or functional redundancy of the microbial mats. To assess these potential changes, we used a network-based analytical framework. To our knowledge, this is the first network-based study that involves a time series approach in freshwater microbial mats.

In our study, we adopted the well-established ecological concepts that ecosystems with a greater number of species are more productive, more resilient to invasions of foreign species, and more stable in the face of perturbations such as drought. Our analysis is based on the pioneering work of David Tilman, whose work has shown that, the greater number of species (specially plants), the larger number of limiting resource interactions are needed to fulfill different needs (Chapin et al. 1998; Tilman 1999, 2004; Tilman et al. 2006). In this general ecological context, almost 10 years ago, Bascompte and Stouffer (2009) suggested that, in face of the ongoing global change crisis, understanding the structure of species relationships, and how that structure relates to network disassembly, should be a high priority for system-level conservation biology (Bascompte et al. 2003; Bascompte 2010; Guimarães et al. 2006, 2017; Rohr et al. 2014). However, in microbial ecology, little is known about the community-wide implications of human-induced perturbations or about the changes that such drivers induce in network interactions among microbial species.

Understanding Microbial Relationships by Network Inference

Ecological interactions have been acknowledged to play a key role in shaping biodiversity (Guimarães et al. 2017). In contrast to ecologists of large organisms who have a long tradition of studying how species interact with each other (Bascompte 2010), microbial ecologists have largely focused on evaluating the relative abundances of community members (taxa and genes) and using various ecological indexes (i.e., Shannon) to measure the degree of resistance, resilience, or functional redundancy following an environmental disturbance (Allison and Martiny 2008; Konopka 2009; Shade et al. 2012). This disparity is likely due to the difficulties in inferring ecological interactions in the microbial world, where, in contrast to eco-logical studies of plant or insect communities, it is very difficult to recognize inter-actions, such as who eats who (antagonism), or mutually beneficial relationships (i.e., a bee pollinating a type of flower) (Bascompte 2010). For microbial communi-ties, it is close to impossible to determine every type of interaction just by observa-tion, as microbial communities are composed of hundreds or thousands of species and exhibit much larger and more rapid changes in biomass, composition, and activ-ity than do plant and animal population (Song et al. 2014). Therefore, it is necessary to use indirect, inferential methods that allow us to predict and detect different types of interactions within complex microbial communities.

Currently, advances in metagenomics and bioinformatics approaches allow the identification and estimation of the relative abundances and functional capabilities of microbes in complex environmental samples. Prediction of the ecological rela-tionships between microorganisms is achieved through network inference, using presence/absence data or abundance profiles generated by several bioinformatics methods. In general, there are two main groups of network inference methods: those that can predict relationships between two species (namely, pairwise relationships) and those that can predict more complex ones (for a review see Faust and Raes 2012; Song et al. 2014).

The predicted relationships are defined as competitive $(-/-)$ (in which two spe-cies consume shared resources) if their abundances across samples are anticorre-lated despite sharing environmental tolerance niches. In contrast, cooperative relationships (where the metabolites produced by one species are consumed by another and, potentially, vice-versa) display similar abundance patterns. Furthermore, the relationships between species can be bidirectional, if the impact is mutually positive (mutualism if obligatory or synergism if nonobligatory), mutually negative (competition), or positive on one side but negative on the other (antagonism); unidi-rectional, if the impact on one of the two is neutral, regardless of whether the impact on the other is positive (commensalism) or negative (amensalism); and nondirec-tional, if the impact on each other is negligible or insignificant (neutralism) (Freilich et al. 2011; Song et al. 2014) (for schematic representation of the ecological interac-tion, see Fig. 7.1 in Faust and Raes 2012 or a more complete description in Table 1 in Song et al. 2014).

The network inference approach also offers potential insight into the stability of microbial communities by using temporal dynamic variation. As reviewed by Faust et al. (2015), several methods have been proposed to infer a single network of inter-action from the entire time series. Existing methods require assuming a particular population dynamics model, which is not known a priori (Xiao et al. 2017). One of these methods exploits a direct comprehensive interacting network for microorgan-isms using Lotka-Volterra equations (Shaw et al. 2016), an approach that has been employed by ecologists to describe a dynamic trophic web of more than two popu-lations of macroorganisms. In microbial ecology studies, it has been observed that Lotka-Volterra equations can quantify microbial interactions and successfully pre-dict microbiome temporal dynamics by modeling changes in abundance as a func-tion of taxon-specific growth rates and pair-wise interaction strengths (Shaw et al. 2016; Stein et al. 2013; Weng et al. 2017).

How Is the Loss of Water Affecting the Microbial Mats Within Churince Lagoon?

Water depletion altered the appearance of the microbial mats of the three sampling sites from two contrasting environments (dry and wet). We analyzed the shifts in community composition, structure, and function of these communities but also the extent to which the specific type of interaction and overall network structure was altered. Our results indicate that, even if the physicochemical environment was constant over time, the taxonomic composition of microbial communities would be continually changing due to the intrinsic dynamics driven by ecological interactions. When environmental influences are not constant (as in the case of water depletion), those fluctuations further affect the community dynamics at metabolic level. Consistent with the altered appearance of wet purple microbial mats, genera that were enriched under dry conditions were taxa implicated in the oxidation of inorganic reduced sulfur compounds. These were the purple sulfur bacterium – Ectothiorhodospiraceae – and the colorless sulfur bacterium *Thioalkalivibrio*, which is physiologically and metabolically adapted to hyper-saline (up to saturation) and alkaline (pH up to 10.5) conditions (Foti et al. 2006; Sorokin et al. 2011). These fluctuations in hydrology in the Lagunita Pond would be expected to have a profound effect on redox potential within microbial mats, due to changes in nutrient and oxygen availability (Peralta et al. 2014). When moist conditions were reestablished by the restoration of the Garabatal wetland (also part of the CCB), the redox gradient was reinstated, allowing the growth of microaerophilic and anaerobic taxa (i.e., increases of sulfate-reducing bacteria) or *Bacteroidetes*. These dynamics indicate that, while few taxa are shared, many unique taxa exist among geographically separated mats in moist conditions. Our results confirm previous results showing that sediments of hyper-saline lakes and lagoons may support a rich community of anaerobic halophilic bacteria, given that the solubility of oxygen in hyper-saline brines is low and supplies of organic matter available are often high (Oren 1988, 2008). Our findings show that these two microbial groups could be used as indicators of water reestablishment due their continuous presence during moist conditions across broad spatial-temporal scales. The first taxon, *Ornatilinea* (*Chloroflexi*), is related to sequenced microbial mats from hot water emerging from a 2775-m deep well (Podosokorskaya et al. 2013), supporting the deepwater origin hypothesis presented by Wolaver et al., (2012) to explain the uniqueness of the CCB springs. The second taxon belongs to an unclassified genus from the *Rikenellaceae* family (*Bacteroidetes*), whose members are recognized as saccharolytic and able to ferment glucose to acid by-products by anaerobic metabolism (Graf 2014).

The similarities among the wet patches during dry conditions (De Anda et al. under revision) suggest a possible role of community members that are able to grow in acidic environments, gaining energy from the oxidation and inorganic sulfur compound and possible ferrous iron to obtain organic carbon from carbon dioxide. We suggest that during water depletion events, small infiltrations of deep water rich in sulfur are patchy, resulting in some wet patches and another dry site. Wet patches possibly contain significant concentrations of sulfide derived from groundwater.

Network Community Structure Indicates Overlapping Niches Within Microbial Mats

We used a Lotka-Volterra-based network inference approach to understand how environmental perturbation affects the stability and dynamics of microbial mats in the Churince system. To distinguish between intrinsic shifts generated by the community per se and those derived from outside forces, we studied microbial mats from three sites within a small pond of the Churince system over the course of 2 years (Fig. 7.3).

Consistent with metabolic coupling within the microbial mats, we found that the overall community structure is densely interconnected, pattern that is interpreted as indicating that the species in the community have overlapping niches (Faust and Raes 2012). This was assessed using a metric known as modularity (Q), which quantifies patterns of connectedness within and across groups (Newman 2006). Positive modularity values indicate that interactions occur predominantly within groups, while negative values show that interactions are more frequent between groups than within them.

Several studies suggest that modularity values for of biological networks fall in the range of 0.3–0.7, with higher values being rare (Newman and Girvan 2004). However, in our study we obtained modularity values close to zero, indicating that, within the microbial mats of Churince, there is a notable lack of modules. Instead, the network of interactions represents "a higher-order module," in which all its members are related with each other (Grilli et al. 2016). This result suggests that the microbial mats of Churince have a unique ecological structure.

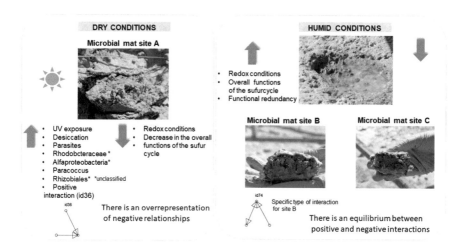

Fig. 7.3 Main results obtained from our time series, Lotka-Volterra-based network inference approach for microbial mats collected during and after overexploitation of water resources within Churince system. Only the microbial mats obtained for our first sampling point (see Fig. 7.1) are shown, a period when Lagunita Pond was dry, and there were only two remaining patches (sites B and C)

Microbial mats are constructed by a complex matrix of trophic and metabolic interactions that is constantly transforming their environment (Kerr et al. 2002). These successful ecological communities have survived millions of years of environmental disturbances due to their ability to construct their own "niche space" (Souza et al., Chapter 6, book 1 of this series). They do so because of metabolic complementation reactions with overlapping ecological functions at a millimeter scale in the absence of algae (that grow too fast and would outcompete other microbes for light) and particular conditions: water, a redox potential given by a sulfur source, and sunlight (Bolhuis et al. 2014; Des Marais 2003; Pajares et al. 2012; Prieto-Barajas et al. 2017; van Gemerden 1993). However, microbial mats grow slowly possibly due to their cohesive development in the Precambrian, perhaps reflecting their evolutionary origins in a world where phosphorus (P) was very limited (Elser et al. 2006) and there was no opportunity to rapidly proliferate. Nowadays, fast-growing communities in a P-rich world have displaced microbial mats. Hence, microbialites like those in CCB constitute "time machines" that exist only in environments in which biogeochemical conditions sustain "multidimensional bubbles" that mimic past conditions, preventing competition by algal growth.

Conclusions

To date, understanding the complex links between environmental disturbance and microbial diversity that affect ecosystem stability has been based on the compositional and functional dynamics of taxa following an environmental disturbance. Understanding microbial interactions is essential for revealing community assembly rules that determine how habitat traits affect the assembly of microbial communities. In this chapter, we highlight that the use of network inference approaches was crucial in understanding how anthropogenic disturbance affects one of the most resistant microbial communities known, microbial mats. These analyses provided some of the first mechanistic insights into the response of microbial mats to environmental perturbation in nearly closed system. Our work indicates that improving the prediction and management of environmental ecosystems in face of anthropogenic perturbations will require further examination of the interaction networks in similar and contrasting environments to improve community prediction under perturbation.

References

Allison SD, Martiny JBH (2008) Resistance, resilience, and redundancy in microbial communities. PNAS 105(Suppl 1):11512–11519
Bascompte J (2010) Structure and dynamics of ecological networks. Science 329:765–766
Bascompte J, Stouffer DB (2009) The assembly and disassembly of ecological networks. Philos Trans R Soc Lond B Biol Sci 364:1781–1787. https://doi.org/10.1098/rstb.2008.0226

Bascompte J, Jordano P, Melián CJ et al (2003) The nested assembly of plant-animal mutualistic networks. PNAS 100:9383–9387. https://doi.org/10.1073/pnas.1633576100

Bolhuis H, Cretoiu MS, Stal LJ (2014) Molecular ecology of microbial mats. FEMS Microbiol Ecol 90:335–350. https://doi.org/10.1111/1574-6941.12408

Carson EW, Souza V, Espinosa-Pérez H et al (2015) Mitochondrial DNA diversity and phylogeography of *Lucania interioris* inform biodiversity conservation in the Cuatro Ciénegas Basin, México. West N Am Nat 75:200–208. https://doi.org/10.3398/064.075.0208

Cerritos R, Eguiarte LE, Avitia M et al (2011) Diversity of culturable thermo-resistant aquatic bacteria along an environmental gradient in Cuatro Ciénegas, Coahuila, México. Antonie Van Leeuwenhoek 99:303–318. https://doi.org/10.1007/s10482-010-9490-9

Chapin FS, Sala OE, Burke IC et al (1998) Ecosystem consequences of changing biodiversity. Bioscience 48:45–52. https://doi.org/10.2307/1313227

Des Marais DJ (2003) Biogeochemistry of hypersaline microbial mats illustrates the dynamics of modern microbial ecosystems and the early evolution of the biosphere. Biol Bull 204:160–167

Elser JJ, Watts J, Schampel JH et al (2006) Early Cambrian food webs on a trophic knife-edge? A hypothesis and preliminary data from a modern stromatolite-based ecosystem. Ecol Lett 9:292–300. https://doi.org/10.1111/j.1461-0248.2005.00873.x

Faust K, Raes J (2012) Microbial interactions: from networks to models. Nat Rev Microbiol 10:538–550. https://doi.org/10.1038/nrmicro2832

Faust K, Lahti L, Gonze D et al (2015) Metagenomics meets time series analysis: unraveling microbial community dynamics. Curr Opin Microbiol 25:56–66. https://doi.org/10.1016/j.mib.2015.04.004

Foti M, Ma S, Sorokin DY et al (2006) Genetic diversity and biogeography of haloalkaliphilic sulphur-oxidizing bacteria belonging to the genus Thioalkalivibrio. FEMS Microbiol Ecol 56:95–101. https://doi.org/10.1111/j.1574-6941.2006.00068.x

Freilich S, Zarecki R, Eilam O et al (2011) Competitive and cooperative metabolic interactions in bacterial communities. Nat Commun 2:587–589. https://doi.org/10.1038/ncomms1597

Graf J (2014) The family Rikenellaceae. In: Rosenberg E, DeLong EF, Lory S et al (eds) The prokaryotes: other major lineages of bacteria and the archaea. Springer, Berlin, Heidelberg, pp 857–859

Grilli J, Rogers T, Allesina S (2016) Modularity and stability in ecological communities. Nat Commun 7:12031. https://doi.org/10.1038/ncomms12031

Guimarães PR, Rico-Gray V, dos Reis SF, Thompson JN (2006) Asymmetries in specialization in ant-plant mutualistic networks. Proc Biol Sci 273:2041–2047. https://doi.org/10.1098/rspb.2006.3548

Guimarães PR, Pires MM, Jordano P et al (2017) Indirect effects drive coevolution in mutualistic networks. Nature 550:511–514. https://doi.org/10.1038/nature24273

Hernández A, Espinosa-Pérez HS, Souza V (2017) Trophic analysis of the fish community in the Ciénega Churince, Cuatro Ciénegas, Coahuila. PeerJ 5:e3637. https://doi.org/10.7717/peerj.3637

Kerr B, Riley MA, Feldman MW, Bohannan BJM (2002) Local dispersal promotes biodiversity in a real-life game of rock-paper-scissors. Nature 418:171–174. https://doi.org/10.1038/nature00823

Konopka A (2009) What is microbial community ecology? ISME J 3:1223–1230. https://doi.org/10.1038/ismej.2009.88

Konopka AE, Lindemann S, Fredrickson JK (2015) Dynamics in microbial communities: unraveling mechanisms to identify principles. ISME J 9:1488–1495. https://doi.org/10.1038/ismej.2014.251

Minckley W (1969) Environments of the Bolson of Cuatro Cienegas, Coahuila, Mexico. Sci Ser 2:1–65

Minckley TA, Jackson ST (2007) Ecological stability in a changing world? Reassessment of the palaeoenvironmental history of Cuatrociénegas, Mexico. J Biogeogr 35:188–190. https://doi.org/10.1111/j.1365-2699.2007.01829.x

Montiel-González C, Tapia-Torres Y, Souza V et al (2017) The response of soil microbial communities to variation in annual precipitation depends on soil nutritional status in an oligotrophic desert. PeerJ 5:e4007. https://doi.org/10.7717/peerj.4007

Newman MEJ (2006) Modularity and community structure in networks. PNAS 103:8577–8582. https://doi.org/10.1073/pnas.0601602103

Newman MEJ, Girvan M (2004) Finding and evaluating community structure in networks. Phys Rev E 69:26113. pmid:14995526

Oren A (1988) Anaerobic degradation of organic compounds at high salt concentrations. Antonie Van Leeuwenhoek 54:267–277. https://doi.org/10.1007/BF00443585

Oren A (2008) Microbial life at high salt concentrations: phylogenetic and metabolic diversity. Saline Systems 4:1–13. https://doi.org/10.1186/1746-1448-4-2

Pajares S, Bonilla-Rosso G, Travisano M et al (2012) Mesocosms of aquatic bacterial communities from the Cuatro Cienegas Basin (Mexico): a tool to test bacterial community response to environmental stress. Microb Ecol 64:346–358. https://doi.org/10.1007/s00248-012-0045-7

Peralta AL, Ludmer S, Matthews JW, Kent AD (2014) Bacterial community response to changes in soil redox potential along a moisture gradient in restored wetlands. Ecol Eng 73:246–253. https://doi.org/10.1016/j.ecoleng.2014.09.047

Podosokorskaya OA, Bonch-Osmolovskaya EA, Novikov AA et al (2013) Ornatilinea apprima gen. nov., sp. nov., a cellulolytic representative of the class Anaerolineae. Int J Syst Evol Microbiol 63:86–92. https://doi.org/10.1099/ijs.0.041012-0

Prieto-Barajas CM, Valencia-Cantero E, Santoyo G (2017) Microbial mat ecosystems: structure types, functional diversity, and biotechnological application. Electron J Biotechnol 31:48–56. https://doi.org/10.1016/j.ejbt.2017.11.001

Rohr RP, Saavedra S, Bascompte J (2014) On the structural stability of mutualistic systems. Science 345:1253497. https://doi.org/10.1126/science.1253497

Shade A, Peter H, Allison SD et al (2012) Fundamentals of microbial community resistance and resilience. Front Microbiol 3:1–19. https://doi.org/10.3389/fmicb.2012.00417

Shaw GTW, Pao YY, Wang D (2016) MetaMIS: a metagenomic microbial interaction simulator based on microbial community profiles. BMC Bioinformatics 17:488. https://doi.org/10.1186/s12859-016-1359-0

Song HS, Cannon W, Beliaev A, Konopka A (2014) Mathematical modeling of microbial community dynamics: a methodological review. Processes 2:711–752. https://doi.org/10.3390/pr2040711

Sorokin DY, Kuenen JG, Muyzer G (2011) The microbial sulfur cycle at extremely haloalkaline conditions of soda lakes. Front Microbiol 2:1–16. https://doi.org/10.3389/fmicb.2011.00044

Souza V, Espinosa-Asuar L, Escalante AE et al (2006) An endangered oasis of aquatic microbial biodiversity in the Chihuahuan desert. PNAS 103:6565–6570. https://doi.org/10.1073/pnas.0601434103

Souza V, Falcón LI, Elser JJ et al (2007) Protecting a window into the ancient earth: towards a Precambrian Park at Cuatro Cienegas, Mexico. The Citizen's page. Evol Ecol Res. Available online at http://www.evolutionary-ecology.com/citizen/citizen.html

Souza V, Siefert JL, Escalante AE et al (2012) The Cuatro Ciénegas Basin in Coahuila, Mexico: an astrobiological Precambrian Park. Astrobiology 12:641–647. https://doi.org/10.1089/ast.2011.0675

Stein RR, Bucci V, Toussaint NC et al (2013) Ecological modeling from time-series inference: insight into dynamics and stability of intestinal microbiota. PLoS Comput Biol 9:31–36. https://doi.org/10.1371/journal.pcbi.1003388

Tilman D (1999) The ecological consequences of changes in biodiversity: a search for general principles. Ecology 80:1455–1474. https://doi.org/10.2307/176540

Tilman D (2004) Niche tradeoffs, neutrality, and community structure: a stochastic theory of resource competition, invasion, and community assembly. Proc Natl Acad Sci 101:10854–10861. https://doi.org/10.1073/pnas.0403458101

Tilman D, Reich PB, Knops JMH (2006) Biodiversity and ecosystem stability in a decade-long grassland experiment. Nature 441:629–632. https://doi.org/10.1038/nature04742

van Gemerden H (1993) Microbial mats: A joint venture. Mar Geol 113:3–25. https://doi.org/10.1016/0025-3227(93)90146-M

Weng FCH, Shaw GTW, Weng CY et al (2017) Inferring microbial interactions in the gut of the Hong Kong whipping frog (Polypedates megacephalus) and a validation using probiotics. Front Microbiol 8:1–11. https://doi.org/10.3389/fmicb.2017.00525

Wolaver BD, Diehl TM (2010) Control of regional structural styles and faulting on Northeast Mexico spring distribution. Environ Earth Sci 62:1535–1549. https://doi.org/10.1007/s12665-010-0639-7

Wolaver BD, Crossey LJ, Karlstrom KE et al (2012) Identifying origins of and pathways for spring waters in a semiarid basin using He, Sr, and C isotopes: Cuatrocienegas Basin, Mexico. Geosphere 9:113–125. https://doi.org/10.1130/GES00849.1

Xiao Y, Angulo MT, Friedman J et al (2017) Mapping the ecological networks of microbial communities. Nat Commun 8:2042. https://doi.org/10.1038/s41467-017-02090-2

Chapter 8
The Magnetotactic Bacteria of the Churince Lagoon at Cuatro Cienegas Basin

Icoquih Zapata-Peñasco, Santiago Bautista-López, and Valeria Souza

Contents

Abstract Magnetotactic bacteria (MTB) are prokaryotes whose movements are directed by the Earth's geomagnetic field. The MTB are diverse in morphology, phylogeny, and physiology. They have unique cellular structures called magnetosomes, which are magnetic mineral crystals (iron) enveloped by a phospholipid bilayer membrane. These magnetosomes confer the ability of bacteria to have magnetotaxis. In this chapter, we will present some findings about the MTB inhabiting Churince Lagoon at Cuatro Cienegas Basin (CCB), such as *Desulfovibrio magneticus*, *Magnetospirillum*, *Magnetospira*, *Magnetococcus*, and *Magnetovibrio*. In a phylogenetic analysis, sequences of genes that encode the magnetosomes from CCB have similarities with those found in marine sediments with volcanic activity. These observations not only conform with other studies that have shown marine ancestry in microbes from CCB but also reaffirms the magmatic influences on the deep aquifer under the Sierra San Marcos and Pinos. Thus, water overexploitation for intensive agriculture in this oasis especially endangers the processes of the iron-sulfur cycle. This biogeochemical cycle is dependent on the deep aquifer and its sediments, which likely function as a depository of ancient anaerobic microbes such as the MTB.

I. Zapata-Peñasco (✉) · S. Bautista-López
Instituto Mexicano del Petróleo, Ciudad de México, México
e-mail: izapata@imp.mx

V. Souza
Departamento de Ecología Evolutiva, Instituto de Ecología, Universidad Nacional Autónoma de México, Ciudad de México, México

© Springer International Publishing AG, part of Springer Nature 2018
F. García-Oliva et al. (eds.), *Ecosystem Ecology and Geochemistry of Cuatro Cienegas*, Cuatro Ciénegas Basin: An Endangered Hyperdiverse Oasis,
https://doi.org/10.1007/978-3-319-95855-2_8

Keywords Magnetotactic bacteria · Sulfur/Iron cycle · Phylogenetic analysis · Marine ancestry · Churince Lagoon

The Iron and Sulfur Cycles

Most of the bodies that comprise the solar system, including the Earth, are composed mainly of iron and silicates. Iron (Fe), because of its density, is at the center of these solar system bodies. The composition of the Earth's core, which has a radius of 3,486 km, is primarily Fe along with a little more than 10% by weight of light elements, such as nickel, copper, and sulfur. The magnetic field of the Earth or geomagnetic field is generated by the iron composition of the nucleus, the convection process that moves the fluid in the outer core, the circulation of electric currents in the atmosphere, and the rotation of the planet. All those forces together make a geo-dynamo effect that is responsible, among other things, for the Van Allen belt that protect us from harmful radiation by the solar wind (Tarbuck et al. 2005). This geomagnetic field is also a compass, a factor in biological evolution that serves as an orientation for many migratory organisms that are particularly sensitive to the magnetic signals. It also influences vital functions inside and outside of all living beings (Binhi and Prato 2017).

The oldest biological processes that have been related to iron are microbial. Consequently, this element is an important cog in Earth's biogeochemical cycles. Early life initiated in the beginning of the Archean Eon, in a warm and ferruginous (anoxic and Fe^{2+} dominated) ocean (Olson and Straub 2016), or, prior to that, in the Hadean Eon ~ 4.1 billion years ago, close to hydrothermal vents (Dodd et al. 2017). In prebiotic Earth within an ancient (Hadean) ocean, it is most likely that H_2S and reduced metal sulfides can remain in solution in anoxic seawater for lengthy periods and could have covered large areas. Thus, life may have originated at highly reduced alkaline submarine springs functioning at a distance from oceanic spreading centers (Russell and Hall 1997).

These submarine hydrothermal vents are found along the mid-ocean ridges where the seafloor spreading occurs. These vents' effluent systems provide seawater with CO_2, H_2S, dissolved H_2, and reduced transition metals, especially iron (Olson and Straub 2016). The theory of a chemoautotrophic origin of life in an iron-sulfur world suggests that primitive autocatalytic metabolism began to produce more complex organic compounds through the catalytic transition of metal centers that became autocatalytic as their organic products turned into ligands. In contrast, carbon fixation metabolism became autocatalytic by forming a metabolic cycle in the form of a primitive sulfur-dependent version of the reductive citric acid cycle (Wächtershäuser 2008). The compartmentalization that is a prior condition for the evolution of any complex system likely involved iron sulfide (FeS) deposited at a warm (< 90 °C) hydrothermal spring, these arrangements may have enabled the movement of molecules of different sizes among compartments; therefore, the last universal common ancestor (LUCA) would account for a set of multiplying, competing, functionally diversifying, and recombining molecules sheltered by compartments (Koonin and Martin 2005). However, traces of the origin of life or even of the diversification of early life in this iron-sulfur-rich ocean are hard to detect. This is why we need analogues of early Earth in order to understand the processes of diversification and coexistence of microbial communities that are an irrefutable early signature of a living planet (Des Marais 2003).

The Extensive Diversity of Magnetotactic Bacteria (MTB)

The MTB are a morphologically, metabolically, and phylogenetically diverse bacterial group capable of incorporating environmental Fe into cell bodies called magnetosomes. The magnetosomes are formed through in vivo synthesis at the nanoscale of magnetic particles such as magnetite (iron oxide, Fe_3O_4) or greigite (iron sulfide, Fe_3S_4). The magnetosomes confer the capacity for magnetotaxis (Bazylinski 1996; Lefèvre and Bazylinski 2013). The MTB have a high potential for diverse applications such as use as biomarkers, specifically paleobiomarkers (because their magnetosomes fossilize in the direction of the poles of Earth), in the synthesis of new nanoparticles and nanomaterials, in bioremediation (i.e., wastewater treatment and heavy metal recovery) and biomedicine (i.e., magnetic resonance imaging, potential cancer treatment, functional magnetosomes for protein display, and gene delivery, magnetic cell separation, biosensor applications, and micro- and nano-manipulators) (Tanaka et al. 2010; Ginet et al. 2011; Mathuriya 2016). Another interesting example of potential use as a bioremediation tool is in the capability of MTB to absorb other toxic metals and metalloids such as tellurium. Moreover, use of these bacteria has emerged as an attractive alternative for the recapture of metals that impact the environment (Tanaka et al. 2010).

Magnetosomes within these particular bacteria are arranged in one or more strings parallel to the major axis of the cell. The lipid bilayer membrane of magnetosome consists mainly of phospholipids, 50% of which are phosphatidylethanolamine (Tanaka et al. 2006). Magnetosome vesicles are formed from the cytoplasmic membrane through an invagination process, in which vesicles are aligned along the cell through a bridge filamentous actin protein. This cellular differentiation process starts with the transport of iron and subsequent crystallization of the magnetite via processes such as nucleation, growth, and morphological regulation (Komeili et al. 2006). It is now known that more than 40 different genes encode magnetosome-associated proteins and yield magnetosome-related phenotypes in *Magnetospirillum* spp. (Uebe and Schüler 2016).

Proteomic analysis indicates that magnetosome proteins are related to the formation of the specialized magnetosome vesicle, biomineralization of iron, and magnetic detection. The genes associated with this complex are encoded into groups of genes within a genomic island, the magnetosome island (MAI). Although this genomic island is large and complex, we can find about 30 *mam* genes (magnetosome membrane) and *mms* (magnetic particle membrane specific) genes that are organized into five polycistronic operons: the large *mam*AB operon (16–17 kb), the small feoAB1 operon, and *mam*GFDC, *mms*6, and *mam*XY operons (Uebe and Schüler 2016). In most cases, genes coding for *Mag*A, *Mms*16, and *Mms*13 exhibit high levels of expression in *Magnetospirillum* (Lohße et al. 2011). For example, the *mms*16 gene encodes GTPase that handles the translocation proteins across cell membranes, as well as the transport of the vesicles within the cell, specifically in the cover assembly of the vesicles. GTPase inhibition prevents synthesis of biomagnetic particles, suggesting that GTPase activity is required for the synthesis of this structure (Okamura et al. 2001; Arakaki et al. 2010). Moreover, the transcription of *mag*A in *Magnetospirillum* is regulated by the concentration of iron in the extracellular environment. The expression of this gene in strains of *E. coli* results in vesicles exhibiting iron storage capacity. Thus, involvement of MagA in iron transport is also implicated in the synthesis of magnetic particles in *Magnetospirillum* (Nakamura et al. 1995; Arakaki et al. 2010). Conversely, if *mms*24 is removed, there is no effect on the invagination of the membrane for the formation of the vesicle. However, the mutated MTB is not able to form crystals of iron oxide, as the accumulation of iron in magnetosome chain is altered since the molecular structure of this protein has the ability to mediate protein-protein interactions and forms multiprotein complexes that regulate the development of magnetosome (Zeytuni et al. 2011).

Although the genomic Island is large and complex, there is evidence of horizontal gene transfer (HGT) playing an important role in the development and distribution of the genes for building magnetosomes in several lineages of prokaryotes. Due to its potential mobility, the MAI is surrounded by mobile elements such as transposases and putative insertion sites near tRNA genes (Lefévre et al. 2013; Uebe and Schüler 2016). Even if there is potential HGT related to this island, comparative phylogenetic analyses of the amino acid sequences of magnetosome proteins and the 16S rRNA gene sequences of various MTB have shown a consistent relationship between the evolution and divergence of these proteins and the 16S rRNA gene, suggesting that the source of magnetotaxis in the *Proteobacteria* phylum is monophyletic (Lefévre et al. 2013).

What Kind of MTB Inhabit Cuatro Cienegas?

Despite the arid climate and its very low phosphorus content, CCB harbors extensive microbial biodiversity. The diverse community includes the living stromatolites and other microbial communities that form the basis of complex food webs (Souza et al. 2012). Due to its high species endemism and its history of evolutionary radiations, CCB is classified as a globally outstanding ecosystem. Furthermore, the microbiota in Cuatro Cienegas have a surprising marine origin (Souza et al. 2006). It is hypothesized that this unique microbiota is due to the deposition of shallow marine limestones and sulfates at the regional uplift, called the Coahuila Island, throughout the late Jurassic to early Cretaceous periods. What is more puzzling is that many of the microbes of CCB can be traced back to a marine origin either in the Jurassic or in the late Precambrian (Moreno-Letelier et al. 2011, 2012). This is also possibly the case for the MTB.

In our study, we sampled sediment of Churince Lagoon at Cuatro Cienegas Basin. Sediment was transported in liquid nitrogen to the laboratory, where we enriched it with synthetic marine water media and imposed a magnetic field in order to select MTB. On the other hand, we isolated environmental DNA from the same sediment and generated clone libraries of three specific genes for magnetosome synthesis, *mms*16, *mms*24, and *mag*A (Arakaki et al. 2010), along with the 16S rRNA gene (Relman 1993). Additionally, the V2–V5 hypervariable regions of 16S rRNA gene were amplified and sequenced by Illumina MiSeq platform.

The 16S rRNA clone analyses from cultured lines (Fig. 8.1) revealed that we retrieved sequences similar to *Magnetospirillum magneticum* (*Alphaproteobacteria* class), originally isolated from freshwater sediment in Japan (Matsunaga et al. 1991), *Desulfovibrio magneticus* (*Deltaproteobacteria* class) (Kawaguchi et al. 1995), *Magnetococcus marinus* (*Alphaproteobacteria* class) (Bazylinski et al. 2013a), and several uncultured *Magnetococcus* clones observed in the Moskva River (Kozyaeva et al. 2017), as well as numerous uncultured bacterium clones obtained from sediment-free enrichment mediating the anaerobic oxidation of methane with sulfate

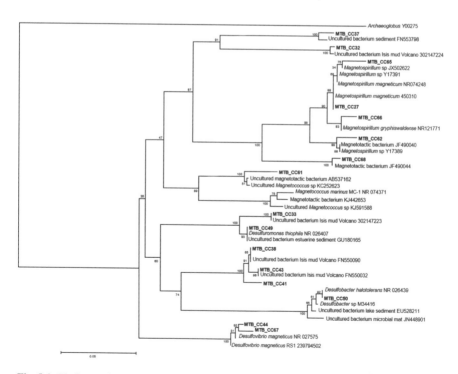

Fig. 8.1 Phylogenetic tree based on retrieved 16S rRNA gene sequence clones (in bold) of MTB from Churince Lagoon. Neighbor-joining analysis with bootstrap of 1000 replicates

from the Isis mud volcano, eastern Mediterranean (Schreiber et al. 2010). With respect to retrieved sequences of magnetosomes, we obtained numerous *mamA* clones similar to *Desulfovibrio magneticus* (Nakazawa et al. 2009), *Magnetospirillum* (Okuda et al. 1996), as well as sequences from marine sediment metagenomes (Yooseph et al. 2007).

The MTB detected in Churince Lagoon through the Illumina 16S RNA gene contributed less than 0.01% of the total abundances, making them a part of the "rare biosphere" (Sogin et al. 2006). The sequences were seen to be similar to *Magnetospirillum*, *Magnetococcus*, *Magnetovibrio*, and *Magnetospira* taxa (Fig. 8.2). At CCB, the rare biosphere is implicated in many microbe-driven processes such as those involved in the sulfur cycle, particularly when environmental conditions are changing (Jousset et al. 2017; Lee et al. 2017). In the microbial mats of the Churince Lagoon, several sulfur keystone taxa have previously been recognized as components of the rare biosphere (De Anda et al. under review).

In another study (Chapter 8 by De Anda et al.), we observed that, in addition to marked local seasonal effects, there is an important anthropogenic effect on microbial biodiversity at CCB due to water overexploitation for intensive agriculture. As the aquifer goes down, diversity declines, especially for the anaerobic microbes that depend on the conditions provided by the deep sulfur- and iron-rich water. This reduction in biodiversity includes, as expected, the MTB guilds (De Anda et al. under review). We hypothesize that MTB will be more resilient to changes than other sulfur-related bacteria because magnetotaxis along with chemotaxis will allow MTB to locate and maintain in vertical chemical concentration gradients in a water column as water declines (Frankel et al. 1997). This vertical alignment may give MTB an advantage in adjusting to environmental alterations. However, this hypothesis remains to be tested.

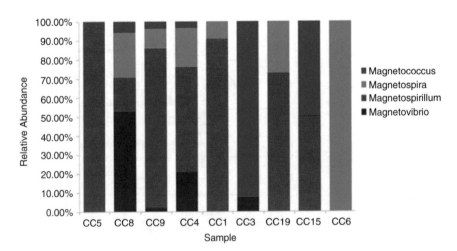

Fig. 8.2 Relative abundance of MTB sequences in sediments of Churince Lagoon, Cuatro Cienegas

The presence of MTB at Churince sediments is not a surprise. These bacteria inhabit sediments of freshwater, brackish, marine, and hypersaline environments. They flourish in chemically stratified water columns at the oxic-anoxic interface, where a redox gradient is established with different compounds that can be used as substrate, such as sulfur compounds (Bazylinski et al. 2013b; Lefèvre and Bazylinski 2013). Furthermore, the distribution, presence, and abundance of MTB are affected by environmental parameters such as salinity, temperature, nitrate, or sulfur compounds (Lin et al. 2013; Liu et al. 2017). As might be expected, the Churince Lagoon is a biogeochemically suitable environment for development of an extensive diversity of MTB. At CCB, and in particular in the Churince system, the sulfur cycle plays a substantial role as a key biogeochemical gear. Previous studies by our group developed MEBS (Multigenomic Entropy Based Score), an algorithm to measure the impact of the sulfur metabolic machinery. These calculations suggested that microbial communities in mats at Cuatro Cienegas have among the highest sulfur scores observed, similar to those for hydrothermal vent and marine benthic communities (De Anda et al. 2017).

Conclusion

The sulfur cycle is a fundamental collection of processes in the development of the structure of microbial communities in Churince Lagoon, a relict oasis in Cuatro Cienegas. The presence of MTB is determined not only by iron but also by other factors, such as sulfur sources. Further studies are required to highlight their ecological function in this desert oasis. Currently, we are studying the whole community of prokaryotes related to MTB in sediments of the Churince Lagoon. These findings will provide us with more information defining the relationships between the important species that assemble the microbial core, the rare biosphere, and keystone taxa of these ancient and diverse communities.

References

Arakaki A, Shibusawa M, Hosokawa M et al (2010) Preparation of genomic DNA from a single species of uncultured magnetotactic bacterium by multiple-displacement amplification. Appl Environ Microbiol 76:1480–1485. https://doi.org/10.1128/AEM.02124-09

Bazylinski DA (1996) Controlled biomineralization of magnetic minerals by magnetotactic bacteria. Chem Geol 132:191–198. https://doi.org/10.1016/S0009-2541(96)00055-1

Bazylinski DA, Williams TJ, Lefévre CT et al (2013a) *Magnetococcus marinus* gen. nov., sp. nov., a marine, magnetotactic bacterium that represents a novel lineage (Magnetococcaceae fam. nov., Magnetococcales ord. nov.) at the base of the Alphaproteobacteria. Int J Syst Evol Microbiol 63:801–808. https://doi.org/10.1099/ijs.0.038927-0

Bazylinski DA, Lefèvre CT, Schüler D (2013b) Magnetotactic bacteria. In: Rosenberg E, EF DL, Lory S et al (eds) The prokaryotes. Springer, Berlin, pp 453–494

Binhi VN, Prato FS (2017) Biological effects of the hypomagnetic field: an analytical review of experiments and theories. PLoS One 12(6):e0179340. https://doi.org/10.1371/journal. pone.0179340

De Anda V, Zapata-Peñasco I, Poot-Hernandez AC et al (2017) MEBS, a software platform to evaluate large (meta)genomic collections according to their metabolic machinery: unravelling the sulfur cycle. GigaScience 6:1–17. https://doi.org/10.1093/gigascience/gix096

Des Marais DJ (2003) Biogeochemistry of hypersaline Microbial mats illustrates the dynamics of the modern microbial ecosystems and early evolution of the biosphere. Biol Bull 204:160–167. https://doi.org/10.2307/1543552

Dodd MS, Papineau D, Grenne T et al (2017) Evidence for early life in Earth's oldest hydrothermal vent precipitates. Nature 543:60–64. https://doi.org/10.1038/nature21377

Frankel RB, Bazylinski DA, Johnson MS, Taylor BL (1997) Magneto-aerotaxis in marine coccoid bacteria. Biophys J 73:994–1000. https://doi.org/10.1016/S0006-3495(97)78132-3

Ginet N, Pardoux R, Adryanczyk G et al (2011) Single-step production of a recyclable nanobio-catalyst for organophosphate pesticides biodegradation using functionalized bacterial magnetosomes. PLoS One 6(6):e21442. https://doi.org/10.1371/journal.pone.0021442

Jousset A, Bienhold C, Chatzinotas A et al (2017) Where less may be more: how the rare biosphere pulls ecosystems strings. ISME J 11:853–862. https://doi.org/10.1038/ismej.2016.174

Kawaguchi R, Burgess JG, Sakaguchi T et al (1995) Phylogenetic analysis of a novel sulfate-reducing magnetic bacterium RS-1, demonstrates its membership of the delta-Proteobacteria. FEMS Microbiol Lett 126:277–282. https://doi.org/10.1111/j.1574-968.1995.tb07430.x

Komeili A, Li Z, Newman DK et al (2006) Magnetosomes are cell membrane invaginations organized by the actin-like protein *Mam*K. Science 311:242–245. https://doi.org/10.1126/science.1123231

Koonin EV, Martin W (2005) On the origin of genomes and cells within inorganic compartments. Trends Genet 21(12):647–654. https://doi.org/10.1016/j.tig.2005.09.006

Kozyaeva VV, Grouzdev DS, Dziuba MV et al (2017) Diversity of magnetotactic bacteria of the Moskva River. Microbiology 86:106–112. https://doi.org/10.1134/S0026261717010088

Lee ZM-P, Poret-Peterson AT, Siefert JL et al (2017) Nutrient stoichiometry shapes microbial community structure in an evaporitic shallow pond. Front Microbiol 8:949. https://doi.org/10.3389/fmicb.2017.00949

Lefèvre DA, Bazylinski A (2013) Ecology, diversity, and evolution of magnetotactic bacteria. Microbiol Mol Biol Rev 77:497–526. https://doi.org/10.1128/MMBR.00021-13

Lefévre CT, Trubitsyn D, Abreu F et al (2013) Monophyletic origin of magnetotaxis and the first magnetosomes. Environ Microbiol 15:2267–2274. https://doi.org/10.1111/1462-2920.12097

Lin W, Wang Y, Gorby Y et al (2013) Integrating niche-based process and spatial process in biogeography of magnetotactic bacteria. Sci Rep 3:1643. https://doi.org/10.1038/srep01643

Liu J, Zhang W, Li X et al (2017) Bacterial community structure and novel species of magnetotactic bacteria in sediments from a seamount in the Mariana volcanic arc. Sci Rep 7:17964. https://doi.org/10.1038/s41598-017-17445-4

Lohße A, Ullrich S, Katzmann E et al (2011) Functional analysis of the magnetosome island in *Magnetospirillum gryphiswaldense*: the mamAB operon is sufficient for magnetite biomineralization. PLoS One 6(10):e25561. https://doi.org/10.1371/journal.pone.0025561

Mathuriya AS (2016) Magnetotactic bacteria: nanodrivers of the future. Crit Rev Biotechnol 36:788–802. https://doi.org/10.3109/07388551.2015.1046810

Matsunaga T, Sakaguchi T, Tadakoro F (1991) Magnetite formation by a magnetic bacterium capable of growing aerobically. Appl Microbiol Biotechnol 35:651. https://doi.org/10.1007/BF00169632

Moreno-Letelier A, Olmedo G, Eguiarte LE, et al (2011) Parallel evolution and horizontal gene transfer of the pst operon in Firmicutes from oligotrophic environments. Int J Evol Biol. Article ID 781642. https://doi.org/10.4061/2011/781642

Moreno-Letelier A, Olmedo-Alvarez G, Eguiarte LE et al (2012) Divergence and phylogeny of Firmicutes from the Cuatro Ciénegas Basin, Mexico: a window to an ancient ocean. Astrobiology 12:674–684. https://doi.org/10.1089/ast.2011.0685

Nakamura C, Kikuchi T, Burgess JG et al (1995) An iron-regulated gene, *mag*A, encoding an iron transport protein of *Magnetospirillum* sp. strain AMB-1. J Biol Chem 270:28392–28396. https://doi.org/10.1074/jbc.270.47.28392

Nakazawa H, Arakaki A, Narita-Yamada S et al (2009) Whole genome sequence of *Desulfovibrio magneticus* strain RS-1 revealed common gene clusters in magnetotactic bacteria. Genome Res 19:1801–1808. https://doi.org/10.1101/gr.088906.108

Okamura Y, Takeyama H, Matsunaga T (2001) A magnetosome- pecific GTPase from the magnetic bacterium *Magnetospirillum magneticum* AMB-1. J Biol Chem 276:48183–48188. https://doi.org/10.1074/jbc.M106408200M100099200

Okuda Y, Denda K, Fukumori Y (1996) Cloning and sequencing of a gene encoding a new member of the tetratricopeptide protein family from magnetosomes of *Magnetospirillum magnetotacticum*. Gene 171:99–102. https://doi.org/10.1016/0378-1119(95)00008-9

Olson KR, Straub KD (2016) The role of hydrogen sulfide in evolution and the evolution of hydrogen sulfide in metabolism and signaling. Physiology 31:60–72. https://doi.org/10.1152/physiol.00024.2015

Relman DA (1993) Universal bacterial 16S rRNA amplification and sequencing. In: Persing DH, Smith TF, Tenover FC, White TJ (eds) Diagnostic molecular microbiology: principles and applications. ASM Press, Washington, DC, pp 489–495

Russell MJ, Hall AJ (1997) The emergence of life from iron monosulphide bubbles at a submarine hydrothermal redox and pH Front. J Geol Soc 154:377–402. https://doi.org/10.1144/gsjgs.154.3.0377

Schreiber L, Holler T, Knittel K et al (2010) Identification of the dominant sulfate-reducing bacterial partner of anaerobic methanotrophs of the ANME-2 clade. Environ Microbiol 12(8):2327–2340. https://doi.org/10.1111/j.1462-2920.2010.02275.x.

Sogin ML, Morrison HG, Huber JA et al (2006) Microbial diversity in the deep sea and the underexplored "rare biosphere". Proc Natl Acad Sci 103:12115–12120. https://doi.org/10.1073/pnas.0605127103

Souza V, Espinosa-Asuar L, Escalante AE et al (2006) An endangered oasis of aquatic microbial biodiversity in the Chihuahuan desert. Proc Natl Acad Sci 103(17):6565–6570. https://doi.org/10.1073/pnas.0601434103

Souza V, Siefert JL, Escalante AE et al (2012) The Cuatro Ciénegas Basin in Coahuila, Mexico: an Astrobiological Precambrian Park. Astrobiology 12:641–647. https://doi.org/10.1089/ast.2011.0675

Tanaka M, Okamura Y, Arakaki A et al (2006) Origin of magnetosome membrane: proteomic analysis of magnetosome membrane and comparison with cytoplasmic membrane. Proteomics 6:5234–5247. https://doi.org/10.1002/pmic.200500887

Tanaka M, Arakaki A, Staniland SS et al (2010) Simultaneously discrete biomineralization of magnetite and tellurium nanocrystals in magnetotactic bacteria. Appl Environ Microbiol 76:5526–5532. https://doi.org/10.1128/AEM.00589-10

Tarbuck EJ, Lutgens FK, Tasa D (2005) Earth: an introduction to Physical Geology, 8th edn. Pearson Education, Inc/Prentice Hall, Upper Saddle River

Uebe R, Schüler D (2016) Magnetosome biogenesis in magnetotactic bacteria. Nat Rev Microbiol 14:621–637. https://doi.org/10.1038/nrmicro.2016.99

Wächtershäuser G (2008) Iron-Sulfur world. In: Wiley encyclopedia of chemical biology, pp 1–8. https://doi.org/10.1002/9780470048672.wecb264

Yooseph S, Sutton G, Rusch DB et al (2007) The Sorcerer II Global Ocean sampling expedition: expanding the universe of protein families. PLoS Biol 5(3):e16. https://doi.org/10.1371/journal.pbio.0050016

Zeytuni N, Ozyamak E, Ben-Harush K et al (2011) Self-recognition mechanism of MamA, a magnetosome-associated TPR-containing protein, promotes complex assembly. Proc Natl Acad Sci 108:E480–E487. https://doi.org/10.1073/pnas.1103367108

Chapter 9
Ecological Adaptability of *Bacillus* to Extreme Oligotrophy in the Cuatro Cienegas Basin

Jorge Valdivia-Anistro, Luis E. Eguiarte, and Valeria Souza

Contents

Abstract The genus *Bacillus* is known for its ability to colonize diverse environments and to take up a wide variety of resources. Both properties are linked to its spore-forming life history strategy and to its high number of *rrn* operon copies per genome. Experimental evidence has postulated a relationship between the number of copies of the *rrn* operon and the availability of environmental phosphorus. Generally, aquatic bacteria isolated from oligotrophic environments have few *rrn* operon copies and other adaptations that decrease cellular phosphorus demand. The Cuatro Cienegas Basin (CCB) is an aquatic ecosystem with extreme oligotrophy and with high diversity of *Bacillus* strains. For this reason, we explored the variation of the *rrn* operon copy number in different *Bacillus* lineages and their physiological implications during growth under oligotrophic conditions. Unexpectedly, the *Bacillus* from the CCB has a high variation in the number of *rrn* operon copies despite the extreme phosphorus limitation in this environment. In addition, these

J. Valdivia-Anistro
Facultad de Estudios Superiores Zaragoza, Universidad Nacional Autónoma de México, Ciudad de México, México

L. E. Eguiarte · V. Souza (✉)
Departamento de Ecología Evolutiva, Instituto de Ecología, Universidad Nacional Autónoma de México, Ciudad de México, México
e-mail: souza@unam.mx

© Springer International Publishing AG, part of Springer Nature 2018
F. García-Oliva et al. (eds.), *Ecosystem Ecology and Geochemistry of Cuatro Cienegas*, Cuatro Ciénegas Basin: An Endangered Hyperdiverse Oasis,
https://doi.org/10.1007/978-3-319-95855-2_9

bacilli showed different ecological responses reflected in the heterogeneity of their growth dynamics. This heterogeneity seems to be a response to the low availability of nutrients and the competitive cost represented by a high number of *rrn* operon copies. Interestingly, the cellular stoichiometry and protein content during growth dynamics of these *Bacillus* are not consistent with the growth rate hypothesis. The ecological adaptability of the genus *Bacillus* to the oligotrophy of the CCB appears to be due to its high heterogeneity in the number of copies of the *rrn* operon, its cellular stoichiometry, and its ecophysiological adaptations.

Keywords Aquatic bacteria · Bacillus · Growth Rate Hypothesis · Phosphorus · *rrn* operon

Introduction

The Chemistry of Life

The Earth's history has been characterized by diverse environmental changes, all of which have had a great influence on the evolution of life. Nevertheless, before of the origin of life, chemical evolution established the principles to life's development and operation. The main evidence of this evolution process is that all life forms have a remarkably similar chemical composition.

The chemistry of life is based on the transformation of six major chemical elements: carbon (C), hydrogen (H), oxygen (O), nitrogen (N), phosphorus (P), and sulfur (S). The abundance of these elements in various environments can impact the chemical composition of life forms. For example, while phosphorus is an important element for life, it is one of the scarcest in the Solar System, and its availability has had a great influence on the evolution of different organisms.

Phosphorus Availability and Its Influence on Bacterial Evolution

The process of stellar nucleosynthesis of phosphorus is an interesting and controversial subject. The nuclear reaction for the synthesis of this element only occurs in massive stars under extreme temperature ($2.0–3.0 \times 10^9$ K; Maciá 2005; Emir et al. 2013), a condition that accounts for its relatively low availability. However, phosphorus abundance estimates are thought to be underestimated (Cescutti et al. 2012). Phosphorus can acquire several chemical states when it is released from massive stars. This chemical versatility is the reason for its importance in the prebiotic reactions that preceded the origin of life (Schwartz 2006; Pasek et al. 2015).

Phosphorus is stored in Earth's crust in different mineral forms (inorganic phosphorus) (Pasek 2008). These minerals are extracted and transformed by organisms and used in different biological processes, generally as organic phosphorus compounds in which P is bound to carbon via phosphomonoester bonds (Maciá 2005).

In aquatic ecosystems, the inorganic fraction is typically found as orthophosphate (PO_4^{3-}), and, under natural conditions, its concentration is low due to biological removal (Jones 2002; Paytan and McLaughlin 2007). The organic fraction of phosphorus is generally in the cellular components of organisms. It has been observed that low phosphorus availability is an environmental pressure that has influenced the evolution of many forms of life (Elser et al. 2000; Jeyasingh and Weider 2007).

Prokaryotes are considered the first and the most diverse life form on Earth (Whitman et al. 1998; Des Marais and Walter 1999). In particular, bacteria are the most abundant organism in aquatic ecosystems (Oren 2004). Oligotrophic conditions of aquatic ecosystems in particular have influenced the evolution and the distribution of microbial communities (Souza et al. 2008; Lauro et al. 2009). Bacteria have developed different adaptations to reduce cellular demands for phosphorus; these include a decrease in the size of its genome (Luo et al. 2011; Sun and Blanchard 2014), replacement of phospholipids in the cell membrane (Carini et al. 2015; Sebastián et al. 2016), and an increase of phosphate acquisition genes (Martiny et al. 2006, 2009).

The Number of Copies of Ribosomal RNA Operon and Phosphorus Availability

An alternative explanation of how bacteria adapt and respond to environmental challenges is in their ecological life history strategies (Green et al. 2008). The ribosomal RNA (*rrn*) operon is a functional trait related to bacterial life history and has been considered a survival strategy related to resource availability and the ability of bacteria to "shift up" to rapid growth when resources become available (Stevenson and Schmidt 2004; Shrestha et al. 2007). The relationships among life history, genome organization, and cellular elemental composition are a focus of research in biological stoichiometry through the "growth rate hypothesis" (GRH) (Elser and Hamilton 2007). The GRH states that, if the environment has a high phosphorus availability, bacteria with a high number of copies of the *rrn* operon will have a faster growth rate and will show a high phosphorus cellular concentration due to increases in the amount of P-rich ribosomal RNA expressed (Elser 2006). In contrast, bacteria that live under oligotrophic conditions will be characterized by few copies of the *rrn* operon (Fegatella et al. 1998; Strehl et al. 1999; Lauro et al. 2009) and have slow growth rates and low concentrations of cellular RNA and P.

The Extreme Oligotrophy of the Cuatro Cienegas Basin and Its Influence in the Bacterial Evolution

The Cuatro Cienegas Basin (CCB) is a unique hydrologic system with an astonishing microbial diversity. Bacterial communities in the CCB live under extreme oligotrophy, and, surprisingly, they are phylogenetically related to marine microorganisms (Souza et al. 2012). The genus *Bacillus* is one of the more abundant and, consequently, the most studied microbes in the CCB.

The genus *Bacillus* is recognized for its ability to colonize different environments because it can use a wide variety of resources and tolerate diverse stress conditions (Feldgarden et al. 2003). Both characteristics are related to its capability to form spores and to its high number of *rrn* operon copies per genome (Yano et al. 2013). In sequenced genomes, this genus has between five (*B. velezensis* CR-502) to 16 (*B. cereus* CMCC 63305) copies of the *rrn* operon (*rrn*DB, Stoddard et al. 2015).

Bacillus coahuilensis m4-4 (CECT 7197) is an endemic and halophilic bacterium isolated from a desiccation lagoon in the CCB (Cerritos et al. 2008). The genome of this *Bacillus* showed the same adaptations to the oligotrophy described in other microbial groups (Alcaraz et al. 2008; Moreno-Letelier et al. 2011). However, *B. coahuilensis* had nine copies of the *rrn* operon, a number much larger than expected for the low phosphorus conditions in the CCB.

How do these bacilli survive under the extreme phosphorus limitation in the CCB? We explored the variation of the *rrn* operon copy number in different *Bacillus* lineages and described their physiological implications during their growth under oligotrophic conditions.

The Number of Copies of the *rrn* Operon in the Bacilli from the CCB

The CCB is composed by diverse aquatic ecosystems, and all of them have unique environmental properties (Souza et al. 2012). We selected different *Bacillus* strains isolated from sediment, water, and the rhizosphere in three sites with different environmental conditions (Fig. 9.1): (a) the Churince system, one of the sites most impacted by water extraction in the basin, which starts in a freshwater spring that flows to a shallow pond and, subsequently, to a desiccated lagoon; (b) the Rio Mesquites, a river site that connects with lateral ponds, all with an imbalanced elemental composition (C:N:P, 900:150:1); and (c) the Pozas Rojas ("Red Pools"), a complex system with fluctuating conditions in salinity and temperature. These isolates formed a phylogenetic tree composed of 18 related groups with strains described in different habitats but principally in marine ecosystems (Fig. 9.1). Three groups stand out: *B.* sp. m2-34 (group II) and *B. coahuilensis* (group V), both only isolated in the CCB, and *B. sonorensis* (group XI) isolated in the Sonoran Desert. In

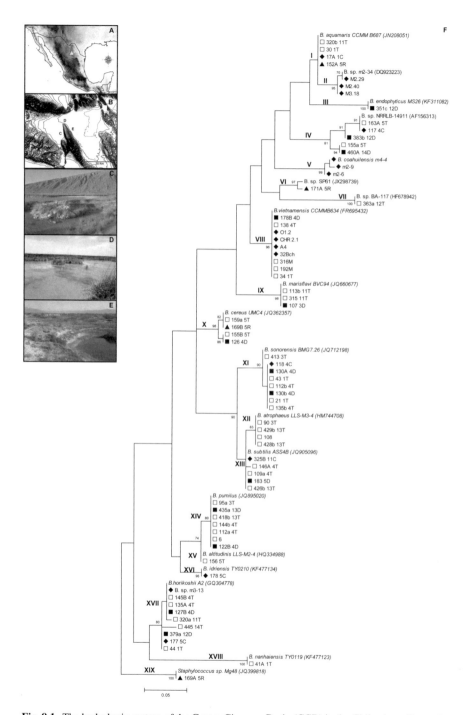

Fig. 9.1 The hydrologic system of the Cuatro Cienegas Basin (CCB) in the Chihuahuan Desert in northeastern México (A, yellow triangle). Location of the principal sites of the isolation of *Bacillus* strains (B): the Churince system (C), the Rio Mesquites (D), and the Pozas Rojas (E). Maximum likelihood (ML) tree of the 18 phylogenetic groups of *Bacillus* (F). Sample type of isolation: □ top section of sediment, ■ bottom section of sediment, ▲ water sediment adjacent to a plant, and ◆ water. (From Valdivia-Anistro et al. 2016)

oligotrophic environments, the coexistence of such high levels of diversity is unexpected; however, in the CCB, this phenomenon may be explained by its extended geological history, variability in environmental conditions, and, above all, too low nutrient availability that enhances evolutionary diversification by dampening horizontal gene flow (Souza et al. 2012).

Considering the importance of oligotrophy in bacterial evolution in the CCB, we quantified the number of copies of the *rrn* operon in every isolate selected. Our prediction was that the extended period of geologic isolation and low phosphorus availability were selective pressures that should lead the decrease of the number of the *rrn* operon copies in these bacilli (Souza et al. 2008; Moreno-Letelier et al. 2011). The isolates showed a range between 6 and 14 *rrn* operon copies (Fig. 9.2). At the time these data were published, six copies are the lowest number quantified in the genus *Bacillus* (Valdivia-Anistro et al. 2016). Surprisingly, this low copy number was observed in the isolates related to the desert strain *B. sonorensis*. However, this number of copies is still relatively high in comparison with other bacteria that live under oligotrophic conditions. Despite this, we consider that this low number of copies could be a response to the environmental conditions and the oligotrophy in the CCB. With respect to the two endemic groups of the CCB, all the isolates of *B.* sp. m2-34 (group II) had eight *rrn* operons, while the *B. coahuilensis* strains (group V) had eight (m4-4), nine (m2-9), and ten (m2-6) copies, respectively (Fig. 9.2). Unexpectedly, we observed a discrepancy between the genomic analysis (nine copies) and the experimental quantification (eight copies) in the type strain of *B. coahuilensis* m4-4. Nevertheless, such discrepancies are not uncommon when

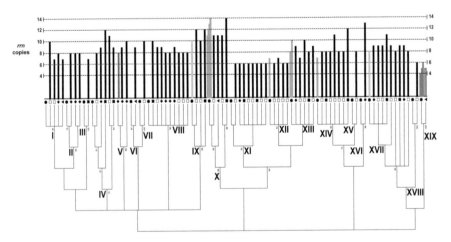

Fig. 9.2 Variability of the rRNA operon copy number among the diversity of *Bacillus* from the CCB. ● type strain of *Bacillus* for every phylogenetic group. Sample type of isolation: □ top section of sediment, ■ bottom section of sediment, ▲ water sediment adjacent to a plant, and ◆ water. The black bars represent the number of operon copies in the isolates from the CCB, and the green bars represent the number of copies in *Bacillus* in the rrnDB. The orange bar represents the number of copies quantified in the type strain of *B. coahuilensis* m4-4 (V). (From Valdivia-Anistro et al. 2016)

these types of data are compared (Vishnivetskaya et al. 2009). Fourteen was the highest copy number observed, found in an isolate in the *B. cereus* group (X) (Fig. 9.2). Fourteen copies of the *rrn* gene are commonly observed in genomic analyses. Overall, our predictions from the GRH were not supported because we expected that the isolates from the CCB to show reduced number of *rrn* operon copies consistent with a slow life history and a K-selected ecological growth strategy.

Previous research on *Bacillus* has focused on the quantification of the number of *rrn* operon copies in different environmental strains (Klappenbach et al. 2000; Shrestha et al. 2007). However, none of these previous studies analyzed the variability in the *rrn* operon copy number in coexisting species of *Bacillus*. We documented that some groups showed intraspecific variation from one to four copies, complicating interpretation of our data (Fig. 9.2). We compared our results against the Ribosomal Database (*rrn*DB, Stoddard et al. 2015). The *rrn*DB compiles every sequenced genome of Bacteria and Archaea and registers the variation in the number of copies of the *rrn* operon. For example, in the *rrn*DB, *B. subtilis* strains showed from eight to ten copies; in our analysis, we observed a somewhat wider range (seven to eleven copies). This variability is common in bacterial genomes. In three species of *Bacillus*, the variation was only from one to three copies (Acinas et al. 2004; Rastogi et al. 2009). The high degree of variability in the number of the *rrn* operon copies in the *Bacillus* strains from the CCB could be an evidence of its adaptation to cope with low phosphorus availability: some strains persist with a K-selected, slow growth rate strategy with low ribosome investment to reduce the cellular demand of this element coexisting with strains with an *r*-selected, fast growth rate strategy with adaptations to take advantage of pulsed resource supply (Alcaraz et al. 2008; Moreno-Letelier et al. 2011).

The Growth Dynamics of *Bacillus* Under Oligotrophic Conditions

In 1949, Jacques Monod wrote the following: "The study of the growth of bacterial cultures does not constitute a specialized subject or branch of research: it is the basic method of Microbiology." For many years, this method was in disuse; however, bacterial growth dynamics is the best mechanism to understand the ecology and physiology of microorganisms (Neidhardt 1999; Schaechter 2006). To understand the physiological implications of living in an oligotrophic environment, we designed an experiment to try to imitate the environmental and nutritional conditions in the CCB. We selected 15 *Bacillus* isolates that represent the variability in the number of the *rrn* operon copies to describe their growth strategies (see Valdivia-Anistro et al. 2016). All the isolates were grown on medium that contained water of the CCB, and the cultures were incubated at the maximum summer water temperature (35 °C). We selected these conditions because the only way to understand the natural growth physiology is to imitate the natural environment (Neidhardt 1999).

The growth dynamics showed high heterogeneity that was uncorrelated with the *rrn* operon copy number. This variability in growth dynamics has been described in other environmental strains with different number of *rrn* operon copies (Dethlefsen and Schmidt 2007). Interestingly, the isolates with the highest number of *rrn* operon copies showed the most extreme values in the growth parameters, as has been seen in other *Bacillus* strains under extreme experimental conditions (Valík et al. 2003; Antolinos et al. 2011, 2012). The heterogeneity in the growth dynamics of the isolates of *Bacillus* could be indicative of the oligotrophic conditions in the CCB and to the competitive costs that are conferred by a high number of *rrn* operon copies. Diverse physiological responses, such as differences in transcription rates or intracellular allocation processes, could be necessary to live in the extreme oligotrophy of the CCB.

The Cellular Stoichiometry of *Bacillus* During the Exponential Phase

Previous studies have described the relationship between the number of *rrn* operon copies, bacterial growth rate, and phosphorus availability (Codon et al. 1995; Elser et al. 2000; Shrestha et al. 2007). In agreement with the GRH, this connection is because the operon number is linked to the synthesis of rRNA which is rich in phosphorus during growth (Elser 2003; Jeyasingh and Weider 2007). We decided to analyze the stoichiometry of *Bacillus* in their nutrient-limited conditions. We collected biomass samples during the exponential growth phase of the *Bacillus* isolates to quantify the content (%) of carbon (C), nitrogen (N), and phosphorus (P). All the isolates showed a variable content of C and N, while the P content was relatively low (0.496%) (Valdivia-Anistro et al. 2016).

Applying the principles of the GRH, we tried to correlate the number of *rrn* operon copies with the growth rate (μ_{max}) and with the cellular phosphorus (P) content. The phosphorus (P) cellular content was not correlated with the growth rate (μ_{max}); however, there was a significant and negative correlation between the numbers of *rrn* operon copies with phosphorus (P) content (Fig. 9.3). This was because the isolates with lower numbers of copies of the operon (B and O) showed the highest phosphorus content and the isolate with the high number of copies (A) had the lowest phosphorus content. The isolates with low operon copies showed low N:P ratios, while isolates with high operon copies had the opposite results (Fig. 9.4). Meanwhile, the *B. coahuilensis* strains (D and E) also had low phosphorus content and intermediate values of this elemental ratio. Interestingly, in comparison with other stoichiometric analysis, the N:P ratios in the *Bacillus* from the CCB were substantially higher than the ratios reported for other bacilli (Loladze and Elser 2011).

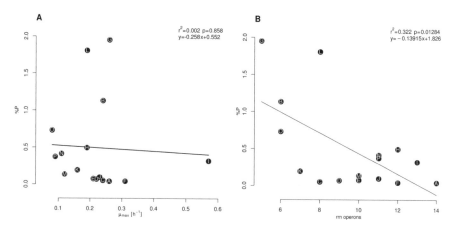

Fig. 9.3 Relationship between the growth rate (μ_{max}) (A) and the number of copies of the *rrn* operon (B), along P (%) content, during the exponential phase. $A = 14$ *rrn* copies, B and O = isolates with 6 *rrn* copies, $D = 9$ and $E = 10$ *rrn* copies (*B. coahuilensis* strains), and $L = 8$ *rrn* copies (From Valdivia-Anistro et al. 2016)

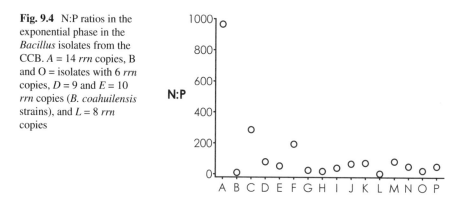

Fig. 9.4 N:P ratios in the exponential phase in the *Bacillus* isolates from the CCB. $A = 14$ *rrn* copies, B and O = isolates with 6 *rrn* copies, $D = 9$ and $E = 10$ *rrn* copies (*B. coahuilensis* strains), and $L = 8$ *rrn* copies

The growth dynamic and the cellular stoichiometry of the *Bacillus* from the CCB were not compatible with the GRH, because the isolates with low copies of the *rrn* operon seem to be more adapted to this oligotrophic this aquatic ecosystem.

Ecophysiological Mechanism in Response to the Oligotrophic Conditions in the CCB

An alternative approach to understanding the effects of oligotrophic conditions of the CCB on the physiology of the genus *Bacillus* is quantification of protein content during growth dynamics. The availability of nutrients plays an important role in protein synthesis during bacterial growth (Dethlefsen and Schmidt 2007; Klumpp et al. 2009; Scott et al. 2010). In the bacilli of the CCB, protein content in every growth phase was variable; however, protein content increased from the lag (λ, 10.11 µg/µl) to the exponential phase (11.80 µg/µl) and decreased in the stationary phase (10.13 µg/µl) (Valdivia-Anistro et al. 2016, unpublished data). This trend of decreasing protein content in the stationary phase is common during bacterial growth (Goelzer and Fromion 2011; Chubukov and Sauer 2014). Some *Bacillus* isolates showed similar patterns of protein content during the growth trajectory (Fig. 9.5). However, these patterns were not related to the number of *rrn* operon copies. Patterns A and B in Fig. 9.5 showed the trend in the protein content commonly seen in other bacteria. However, the *B. coahuilensis* strains have unusual patterns (E and F) in which the protein content increases gradually from the lag to the stationary phase (Fig. 9.5).

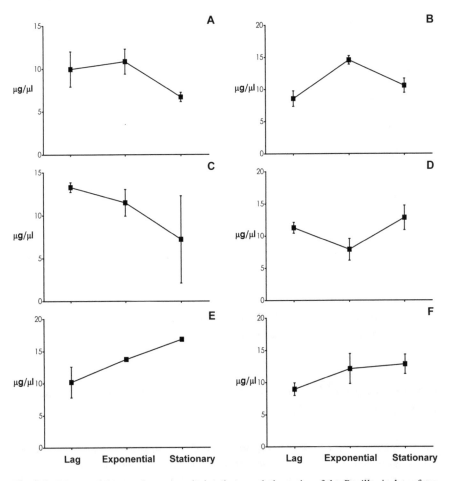

Fig. 9.5 Patterns of the protein content during the growth dynamics of the *Bacillus* isolates from the CCB. Number of copies of the *rrn* operon per pattern: A = 6, 7, 8, 13, and 14; B = 10, 11, and 12; C = 11 (two isolates); D = 6 and 12; E = 8 and 10; and F = 9 (Data not published)

Previously it has been suggested that maximum growth rate reflects levels of nutrients available during protein synthesis (Scott and Hwa 2011). However, protein content was so variable that it showed low relationship with the growth parameters estimated. The isolate with the fastest growth rate did not have the highest protein content, and conversely, some slow-growing isolates had high protein content. This could be important evidence of how these bacilli adapt to oligotrophy in the CCB. Bacteria with fast growth rates are more sensitive to decreased production of stress resistance proteins, a trade-off between growth performance and survival (Goelzer and Fromion 2011).

These discrepancies could be explained by the complexity and variability in the natural environments experienced in the CCB as opposed to the more stable laboratory conditions (Neidhardt 1999; Schaechter 2006). It is possible that, under some environmental conditions, the growth rate is not dependent on nutrient availability. The physiological state and the allocation of resources are processes that influence protein synthesis and limit bacterial growth (Scott et al. 2010; Scott and Hwa 2011). Thus, any physiological mechanism that saves resources or decreases the production of useless proteins could improve bacterial performance. For example, in a minimal medium, *B. subtilis* strains showed a faster growth rate when they lost genes involved in flagella and spore production (Goelzer and Fromion 2011). A similar mechanism is described in the endemic species from the CCB, *B. coahuilensis* (Alcaraz et al. 2008, 2010).

Microbial evolutionary ecology is a complex research topic because it involves multivariable processes. The ecological versatility of the genus *Bacillus* makes it interesting to study but also makes it highly challenging. Biological stoichiometry is a novel and integrative perspective to understand ecology and evolution of life. The GRH connects elemental composition with evolutionary changes in organism growth, all bound to the genetic composition and transcription capacity of the ribosomal RNA genes (Elser and Hamilton 2007). For this reason, the analysis of the variability of the *rrn* operon copy number in *Bacillus* strains isolated from extremely oligotrophic aquatic ecosystems was a promising way to evaluate any association between the variation of this trait and bacterial physiology. This study showed that the *rrn* operon copy number is a variable trait among field-isolated bacilli and this variation is reflected in their growth rate properties and chemical composition. However, we found no simple answers regarding variations in *rrn* operon copy number as a response of this bacterium to a low-nutrient environment. Nevertheless, we confirmed that the number of *rrn* operon copies is trait-related with the evolutionary history of the genus *Bacillus* that reflects its ecological versatility and adaptability to different environmental conditions. Much more complicated was the result that the variability in the *rrn* operon copy number in the *Bacillus* isolated from the CCB does not seem to conform to the broader context of the GRH. Thus, we suggest that the variability of the *rrn* operon, in the cellular stoichiometry and in the ecophysiological mechanisms, described in these bacilli reflect their ecological adaptability to the extreme oligotrophy of the Cuatro Cienegas Basin.

References

Acinas SG, Marcelino LA, Klepac-Ceraj V et al (2004) Divergence and redundancy of 16S rRNA sequences in genomes with multiple *rrn* operons. J Bacteriol 186:2629–2635

Alcaraz LD, Olmedo G, Bonilla G et al (2008) The genome of *Bacillus coahuilensis* reveals adaptations essential for survival in the relic of an ancient marine environment. PNAS 105:5803–5808. https://doi.org/10.1073/pnas.0800981105

Alcaraz LD, Moreno-Hagelsieb G, Eguiarte LE et al (2010) Understanding the evolutionary relationships and major traits of *Bacillus* through comparative genomics. BMC Genomics 11:332. https://doi.org/10.1186/1471-2164-11-332

Antolinos V, Muñoz M, Ros-Chumillas M et al (2011) Combined effect of lysozyme and nisin at different incubation temperature and mild heat treatment on the probability of time growth of *Bacillus cereus*. Food Microbiol 28:305–310. https://doi.org/10.1016/j.fm.2010.07.021

Antolinos V, Muñoz-Cuevas M, Ros-Chumillas M et al (2012) Modelling the effects of temperature and osmotic shifts on the growth kinetics of *Bacillus weihenstephanensis* in broth and food products. Int J Food Microbiol 158:36–41. https://doi.org/10.1016/j.ijfoodmicro.2012.06.017

Carini P, Van Mooy BA, Thrash JC et al (2015) SAR11 lipid renovation in response to phosphate starvation. PNAS 112:7767–7772. https://doi.org/10.1073/pnas.1505034112

Cerritos R, Vinuesa P, Eguiarte LE et al (2008) *Bacillus coahuilensis* sp. nov., a moderately halophilic species from a desiccated lagoon in the Cuatro Cienegas valley in Coahuila, Mexico. Int J Syst Evol Microbiol 58:919–923. https://doi.org/10.1099/ijs.0.64959-0

Cescutti G, Matteucci F, Caffau E et al (2012) Chemical evolution of the Milky Way: the origin of phosphorus. Astron Astrophys 540. https://doi.org/10.1051/0004-6361/201118188

Chubukov V, Sauer U (2014) Environmental dependence of stationary-phase metabolism in *Bacillus subtilis* and *Escherichia coli*. Appl Environ Microbiol 80:2901–2909. https://doi.org/10.1128/AEM.00061-14

Codon C, Liveris D, Squires C et al (1995) rRNA operon multiplicity in *Escherichia coli* and the physiological implications of *rrn* inactivation. J Bacteriol 177:4152–4156

Des Marais DJ, Walter MR (1999) Astrobiology: exploring the origins, evolution, and distribution of life in the universe. Annu Rev Ecol Syst 30:397–420

Dethlefsen L, Schmidt TM (2007) Performance of the translational apparatus varies with the ecological strategies of bacteria. J Bacteriol 189:3237–3245

Elser JJ (2003) Biological stoichiometry: a theoretical framework connecting ecosystem ecology, evolution, and biochemistry for application in astrobiology. Int J Astrobiology 2, 185–193. https://doi.org/10.1017/S1473550403001563

Elser JJ (2006) Biological stoichiometry: a chemical bridge between ecosystem ecology and evolutionary biology. Am Nat 168(Suppl. 6):S25–S35

Elser JJ, Hamilton A (2007) Stoichiometry and the New Biology: the future is now. PLoS Biol 5(7):e181. https://doi.org/10.1371/journal.pbio.0050181

Elser JJ, Sterner RW, Gorokhova E et al (2000) Biological stoichiometry from genes to ecosystems. Ecol Lett 3:540–550

Emir R, Yusof N, Patermann I et al (2013) On the nucleosynthesis of phosphorus in massive stars. AIP Conf Proc 1528:62. https://doi.org/10.1063/1.4803569

Fegatella F, Lim J, Kjelleberg S et al (1998) Implications of rRNA operon copy number and ribosome content in the marine oligotrophic ultramicrobacterium *Sphingomonas* sp. strain RB2256. Appl Environ Microbiol 64:4433–4438

Feldgarden M, Byrd N, Cohan FM (2003) Gradual evolution in bacteria: evidence from *Bacillus* systematics. Microbiology 149:3565–3573

Goelzer A, Fromion V (2011) Bacterial growth rate reflects a bottleneck in resource allocation. Biochim Biophys Acta 1810:978–988. https://doi.org/10.1016/j.bbagen.2011.05.014

Green JL, Bohannan BJM, Whitaker RJ (2008) Microbial biogeography: from taxonomy to traits. Science 320:039–1042. https://doi.org/10.1126/science.1153475

Jeyasingh PD, Weider LJ (2007) Fundamental links between genes and elements: evolutionary implications of ecological stoichiometry. Mol Ecol 16:4649–4661. https://doi.org/10.1111/j.1365-294X.2007.03558.x

Jones RD (2002) Phosphorus cycling. In: Hurst CJ, Crawford RL, Knudsen GR et al (eds) Manual of environmental microbiology. ASM Press, Washington, DC, pp 450–455

Klappenbach JA, Dunbar JM, Schmidt TM (2000) rRNA operon copy number reflects ecological strategies of bacteria. Appl Environ Microbiol 66:1328–1333

Klumpp S, Zhang Z, Hwa T (2009) Growth rate-dependent global effects on gene expression in bacteria. Cell 139:1366–1375. https://doi.org/10.1016/j.cell.2009.12.001

Lauro FM, McDougald D, Thomas T et al (2009) The genomic basis of trophic strategy in marine bacteria. PNAS 106:5527–15533. https://doi.org/10.1073/pnas.0903507106

Loladze I, Elser JJ (2011) The origins of the Redfield nitrogen-to-phosphorus ratio are in a homoeostatic protein-to-rRNA ratio. Ecol Lett 14:244–250. https://doi.org/10.1111/j.1461-0248.2010.01577.x

Luo H, Friedman R, Tang J et al (2011) Genome reduction by deletions of paralogs in the marine cyanobacterium *Prochlorococcus*. Mol Biol Evol 28:2751–2760. https://doi.org/10.1093/molbev/msr081

Maciá E (2005) The role of phosphorus in chemical evolution. Chem Soc Rev 34:691–701

Martiny AC, Coleman ML, Chisholm SW (2006) Phosphate acquisition genes in *Prochlorococcus* ecotypes: evidence for genome-wide adaptation. PNAS 103:12552–12557

Martiny AC, Huang Y, Li W (2009) Occurrence of phosphate acquisition genes in *Prochlorococcus* cells from different ocean regions. Environ Microbiol 11:1340–1347. https://doi.org/10.1111/j.1462-2920.2009.01860.x

Moreno-Letelier A, Olemdo G, Eguiarte LE et al (2011) Parallel evolution and horizontal gene transfer of the *pst* operon in Firmicutes from oligotrophic environments. Int J Evol Biol 2011:781642. https://doi.org/10.4061/2011/781642

Neidhardt FC (1999) Bacterial growth: constant obsession with *dN/dt*. J Bacteriol 181:7405–7408

Oren A (2004) Prokaryote diversity and taxonomy: current status and future challenges. Philos Trans R Soc B Biol Sci 359:623–638

Pasek MA (2008) Rethinking early Earth phosphorus geochemistry. PNAS 105:853–858. https://doi.org/10.1073/pnas.0708205105

Pasek MA, Herschy B, Kee TP (2015) Phosphorus: a case for mineral-organic reactions in prebiotic chemistry. Orig Life Evol Biosph 45:207–218. https://doi.org/10.1007/s11084-015-9420-y

Paytan A, McLaughlin K (2007) The oceanic phosphorus cycle. Chem Rev 107:563–576

Rastogi R, Wu M, Dasgupta I et al (2009) Visualization of ribosomal RNA operon copy number distribution. BMC Microbiol 9:208. https://doi.org/10.1186/1471-2180-9-208

Schaechter M (2006) From growth physiology to system biology. Int Microbiol 9:157–161

Schwartz AW (2006) Phosphorus in prebiotic chemistry. Philos Trans R Soc Lond B Biol Sci 361:1743–1749; discussion 1749

Scott M, Hwa T (2011) Bacterial growth laws and their applications. Curr Opin Biotechnol 22:559–565. https://doi.org/10.1016/j.copbio.2011.04.014

Scott M, Gunderson CW, Mateescu EM et al (2010) Interdependence of cell growth and gene expression: origins and consequences. Science 330:1099–1102. https://doi.org/10.1126/science.1192588

Sebastián M, Smith AF, González JM et al (2016) Lipid remodelling is a widespread strategy in marine heterotrophic bacteria upon phosphorus deficiency. ISME J 10:968–978. https://doi.org/10.1038/ismej.2015.172

Shrestha PM, Noll M, Liesack W (2007) Phylogenetic identity, growth-response time and rRNA operon copy number of soil bacteria indicate different stages of community succession. Enviro Microbiol 9:2464–2474

Souza V, Eguiarte LE, Siefert JS et al (2008) Microbial endemism: does phosphorus limitation enhance speciation? Nat Rev Microbiol 6:559–564. https://doi.org/10.1038/nrmicro1917

Souza V, Siefert JL, Escalante AE et al (2012) The Cuatro Ciénegas Basin in Coahuila, Mexico: an astrobiological Precambrian park. Astrobiology 12:641–647. https://doi.org/10.1089/ast.2011.0675

Stevenson BS, Schmidt TM (2004) Life history implications of rRNA gene copy number in *Escherichia coli*. Appl Environ Microbiol 70:6670–6677

Stoddard SF, Smith BJ, Hein R et al (2015) *rrn*DB: improved tools for interpreting rRNA gene abundance in bacteria and archaea and a new foundation for future development. Nucleic Acids Res 43:D593–D598. https://doi.org/10.1093/nar/gku1201

Strehl B, Holtzendorff J, Partensky F et al (1999) A small and compact genome in the marine cyanobacterium *Prochlorococcus marinus* CCMP 1375: lack of an intron in the gene for tRNAr (Leu)^{UAA} and a single copy of the rRNAop. FEMS Microbiol Lett 181:261–266

Sun Z, Blanchard JL (2014) Strong genome-wide selection early in the evolution of *Prochlorococcus* resulted in a reduced genome through the loss of a large number of small effect genes. PLoS One 9(3):e88837. https://doi.org/10.1371/journal.pone.0088837

Valdivia-Anistro JA, Eguiarte-Fruns LE, Delgado-Sapién G et al (2016) Variability of rRNA operon copy number and growth rate dynamics of *Bacillus* isolated from an extremely oligotrophic aquatic ecosystem. Front Microbiol 6:1486. https://doi.org/10.3389/fmicb.2015.01486

Valík L, Görner F, Lauková D (2003) Growth dynamics of *Bacillus cereus* and self-life of pasteurised milk. Czech J Food Sci 21:195–202

Vishnivetskaya TA, Kathariou S, Tiedje JM (2009) The *Exiguobacterium* genus: biodiversity and biogeography. Extremophiles 13:541–555. https://doi.org/10.1007/s00792-009-0243-5

Whitman WB, Coleman DC, Wiebe WJ (1998) Prokaryotes: the unseen majority. PNAS 95:6578–6583

Yano K, Wada T, Suzuki S et al (2013) Multiple rRNA operons are essential for efficient cell growth and sporulation as well as outgrowth in *Bacillus subtilis*. Microbiology 159:2225–2236. https://doi.org/10.1099/mic.0.067025-0

Chapter 10
Bacterial Siderophore-Mediated Iron Acquisition in Cuatro Cienegas Basin: A Complex Community Interplay Made Simpler in the Light of Evolutionary Genomics

H. Ramos-Aboites, A. Yáñez-Olvera, and F. Barona-Gómez

Contents

Abstract Ferric iron (Fe^{3+}) became abundant after the oxidation event that occurred during the Precambrian, but biologically limited due to its poor and uneven distribution in its soluble form, ferrous iron (Fe^{2+}). In consequence, siderophores, i.e., specialized iron scavenger metabolites, evolved to allow bacteria to obtain this nutrient. Therefore, siderophores can mediate complex bacterial communities, emphasizing the ecological role of these specialized metabolites. In this chapter, we present what is known about hydroxamate siderophores and, in particular, about coelichelin and desferrioxamines that are produced by genera belonging to the phylum *Actinobacteria*. Given that this phylum is predominant in Cuatro Cienegas Basin (CCB), our interest is in the evolution and ecological roles of these specialized metabolites in this unique ecological niche. We review the biosynthetic and transport capabilities sustaining bacterial hydroxamate siderophore-mediated iron acquisition in *Actinobacteria* and provide an example to illustrate a proposed evolutionary conceptual framework

H. Ramos-Aboites · A. Yáñez-Olvera · F. Barona-Gómez (✉)
Unidad de Genómica Avanzada (Langebio), Centro de Investigación y de Estudios
Avanzados del IPN (Cinvestav), Irapuato, México
e-mail: hilda.ramos@cinvestav.mx; alan.yanez@cinvestav.mx;
francisco.barona@cinvestav.mx

© Springer International Publishing AG, part of Springer Nature 2018
F. García-Oliva et al. (eds.), *Ecosystem Ecology and Geochemistry of Cuatro
Cienegas*, Cuatro Ciénegas Basin: An Endangered Hyperdiverse Oasis,
https://doi.org/10.1007/978-3-319-95855-2_10

useful for molecular functional and ecological analyses. The example presented includes genomic analysis of novel actinobacteria that were isolated from CCB that leads to novel biological insights, informing us about the structure and function of the microbial community as mediated by hydroxamate siderophores.

Keywords Desferrioxamines · Coelichelin · Siderophore biosynthesis · Siderophore-binding proteins · *Actinobacteria* · *Lentzea*

Introduction to the Siderophores

Iron is an essential nutrient for virtually all forms of life. It composes up to 32% of the Earth's crust and plays an important role in the metabolism of all living organisms. Iron is required in many fundamental metabolic processes, including the tricarboxylic acid cycle, electron transport chains, oxidative phosphorylation, and photosynthesis (Fardeau et al. 2011) as well as in the biosynthesis of porphyrins, vitamins, antibiotics, toxins, cytochromes, siderophores, pigments, aromatic compounds, and nucleic acids (Messenger and Barclay 1983). It has also been shown that iron plays an important role in the formation of microbial biofilms (as it regulates the surface motility of microorganisms; Saha et al. 2016), in bacterial pathogenesis (Cornelis and Andrews 2010), and in the establishment of ecological relationships (Boiteau et al. 2016). However, the bioavailability of iron in aqueous environments, such as soil, is low: despite being quantitatively abundant in its ferric form (Fe^{3+}), its bioavailability is limited as ferric iron is strongly bound and not soluble in water.

It is therefore assumed that early microorganisms were able to use soluble ferrous iron (Fe^{2+}), which was abundant due to an oxygen-poor atmosphere. However, as oxygen-rich conditions arose, ferrous iron was oxidized to insoluble ferric iron (Fe^{3+}), removing an easily bioavailable source of iron. To survive and respond to this challenge, microorganisms evolved the production of siderophores, which can be broadly defined as organic compounds with low molecular masses that can tightly chelate ferric iron (Fe^{3+}) (Holden and Bachman 2015). These specialized iron scavengers in turn allowed bacteria, fungi, and plants to assimilate essential iron from the environment (Hider and Kong 2010). Once ferric iron-siderophore complexes are formed, they are transported into the cytosol, where the ferric iron gets reduced to ferrous iron, becoming accessible to the various metabolic processes in which this ion plays crucial roles (e.g., catalysis).

The transport system used by bacteria is different for Gram-positive and Gram-negative bacteria. Irrespective of these differences, the ferric iron-siderophore complexes are bound by periplasmic siderophore-binding proteins (SBPs), which are anchored to the bacterial cell membrane. Once the ferric iron-siderophore complexes are recognized outside the cell, these are mobilized into the cytoplasm via ATP-dependent transporters or ABC transport systems (Ahmed and Holmström 2014; Saha et al. 2016). Once the siderophore is bound to ferric iron, the newly formed complex can sit in the environment until it is moved into the cytosol, where the ferric iron gets reduced to its ferrous form, freeing the metal ion from the sidero-phore. After the release of iron, siderophores can either get degraded or recycled by excretion through efflux pump systems (Saha et al. 2016).

More than 500 different types of siderophores are known, of which 270 have been structurally characterized (Ahmed and Holmstrom 2014). Depending on the chemical nature of the moieties donating the oxygen ligand for coordination of Fe^{3+}, siderophores can be classified (Miethke and Marahiel 2007) into three main catego-ries: (i) *carboxylates*, which bind to iron through carboxyl and hydroxyl groups; (ii) *catecholates*, in which each catecholate group supplies two oxygen atoms to form a hexadentate octahedral complex, providing a strong binding even at very low con-centrations of the ion; and (iii) *hydroxamates*, the most common group of sidero-phores found in nature, which consist of C(=O) N-(OH) R, where R is either an amino acid or a derivative of it.

The strong binding between ferric iron and the siderophore is due to oxygen molecules that form a complex with iron. This is particularly the case for hydroxa-mate siderophores, where a bidentate ligand with oxygen atoms is formed. Thus, each siderophore is capable of forming a stable hexadentate octahedral complex with Fe^{3+}. The relevance of this non-covalent binding is that it provides a means to protect the iron-ligand complex against hydrolysis and enzymatic degradation in the environment. Moreover, due to the physicochemical features of the coordination bonds involved during chelation of iron, all siderophores form colorful complexes that can be easily detected and analyzed by spectrophotometric techniques, includ-ing visual inspection (*O*-CAS assay; Schwyn and Neilands 1987; Louden et al. 2011) or liquid chromatography coupled with mass spectrometry (LC-MS) (Saha et al. 2016). This brings about a unique opportunity to analyze the role of these specialized metabolites in the environment and within complex samples in a rela-tively simple fashion.

A canonical example of the hydroxamate siderophores are the desferrioxamines (dFOs), also referred to as ferrioxamines (FOs) when bound to Fe^{3+}. These special-ized metabolites are mainly produced by genera belonging to the phylum *Actinobacteria* (Roberts et al. 2012; Wang et al. 2014; Cruz-Morales et al. 2017), which includes the genus *Streptomyces* but also other unrelated families belonging to other phyla, such as *Gammaproteobacteria* (Essen et al. 2007) and *Alphaproteobacteria* (Smits and Duffy 2011). Despite their broad occurrence, it is not until very recently that these metabolites have been intensively investigated beyond their production and isolation from bacterial cultures. The renewed interest in dFOs relates to the possibilities opened by genomic studies deciphering the bio-synthetic, regulatory, and transport mechanisms that sustain siderophore-mediated iron acquisition (Barona-Gómez et al. 2004, 2006; Kadi et al. 2007; Tierrafría et al. 2011; Cruz-Morales et al. 2017; Ronan et al. 2018). Integration and discussion of what we have learned about dFOs since the beginning of the genomics era, specifi-cally in the context of the CCB, represent the aim of this book chapter.

The Importance of Being Desferrioxamines

The genetic basis for the biosynthesis of hydroxamate siderophores, and its regulation, is relatively well-documented in the model genus *Streptomyces* (Barona-Gómez et al. 2004, 2006; Tunca et al. 2007, 2009; Cruz-Morales et al. 2017). Species of this genus, as well as other closely related actinobacteria, produce different hydroxamate siderophores through divergent biosynthetic gene clusters (BGC). The known pathways involve either non-ribosomal peptide synthetases (NRPS), as in the case of the siderophore coelichelin (Lautru et al. 2005), or NRPS-independent peptide synthetases (NIS), which are responsible for the synthesis of dFOs. Historically, the main dFOs considered to be produced by *Streptomyces* species were the cyclic and the acyl-capped linear forms, dFO-E and dFO-B, respectively (Barona-Gomez et al. 2004, 2006; Yamanaka et al. 2005). Recently, however, novel aryl-capped versions have been reported (Cruz-Morales et al. 2017), as well as a broad range of novel dFOs that were discovered as a surprise after subtly modifying the medium in which *Streptomyces chartreusis* was grown (Senges et al. 2018). Despite this chemical diversity, the biosynthesis of dFOs is well-defined and includes only five enzymes (DesA, DesB, DesC, DesD, and DesG) needed to convert lysine (or ornithine) into monohydroxamates, which in turn are condensed by the NIS enzyme DesD, giving rise to the final product(s) (Fig. 10.1).

With the genetic basis of dFOs biosynthesis on hand, deciphering some aspects of the biochemistry of these specialized metabolites became possible (Kadi et al. 2007; Codd et al. 2018; Ronan et al. 2018). Interestingly, the intrinsic promiscuous nature of the Des enzymes, which seems a relevant ecological feature as argued later in this chapter, has been successfully exploited for the production of novel analogs of dFO-B, as reviewed by Codd and co-workers (2018). In addition to showing the chemical diversity associated with the Des enzymes, these studies are important because the mesylate salt of dFO-B, or desferal (due to its extraordinary properties as an iron chelator), is used to treat patients with secondary iron overload disease or acute iron poisoning. Overall, the approaches followed to obtain these novel potential drugs include feeding of diverse precursors with chemical modifications with the expectation of improving desirable pharmacological characteristics. Indirectly, these experimental schemes have a bearing on the nature of the siderophore systems that have evolved under real environmental conditions (e.g. Gubbens et al. 2017).

Stemming from these genetic and biochemical studies, dFOs have begun to catch the attention of microbial ecologists, as their chemical features may relate to potential mechanisms by which bacteria may evolve and interact within communities (Cruz-Morales et al. 2017; Bruns et al. 2018). Recent microbiological work demonstrates, at least in the laboratory using model bacterial strains, the potential role of dFOs in mediating interactions (Yamanaka et al. 2005; Traxler et al. 2012, 2013; Arias et al. 2015; Galet et al. 2015). Indeed, these microbiological challenges using model actinobacterial organisms have led to dFOs structural diversification (Traxler et al. 2013; Traxler and Kolter 2015). Overall, these studies show how dFOs production and their chemical diversity, which can be overwhelming as recently shown in

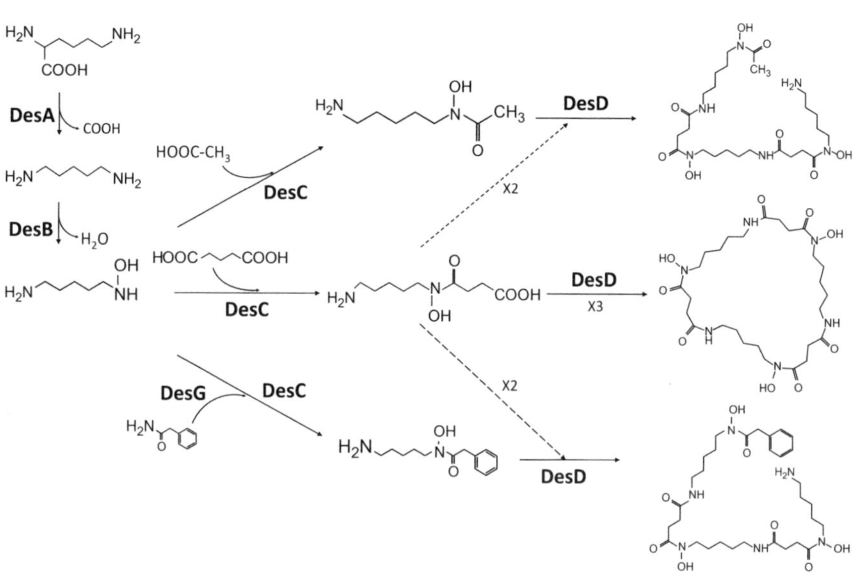

Fig. 10.1 Biosynthetic pathway leading to different dFOs. The biosynthetic scheme is based on five Des enzymes, including the proposal of Cruz-Morales et al. (2017) for the incorporation of phenylacetate by DesG, a homolog of the penicillin amidase, to produce aryl-desferrioxamines

Streptomyces chartreusis (Senges et al. 2018), are prompted by the presence of other bacteria growing in the same culture.

The abovementioned reports explain early work showing that uncultivated bacteria could be isolated after growth induction by dFO-producing bacteria, which turned out to belong to the actinobacterial genus *Micrococcus* (D'onofrio et al. 2010). Indeed, the authors of that study concluded that these organisms allowed siderophore-deficient microorganisms to grow under challenging conditions. In the basis of these observations, and because these properties fulfill recently suggested criteria for the establishment of model ecological systems to investigate specialized metabolites (Pessotti et al. 2018), we propose that *Streptomyces* and related genera from CCB are a suitable biological model for studying the evolution and ecology of dFOs. As a first step toward this possibility, we have previously reported on the first ecological genomic characterization of *Actinobacteria* and their siderophores in CCB (Cruz-Morales et al. 2017).

Regulation of Desferrioxamines and Morphological Development

Gene expression of proteins involved in iron metabolism, such as those directing the synthesis and assimilation of siderophores, is regulated by the action of a family of divalent metal-dependent regulatory proteins, or DmdR repressors, that recognize a specific sequence motif that is situated upstream of the targeted genes, within the promoter or operator regions of iron-related operons (Tunca et al. 2007, 2009). The relationship between this DNA sequence (5'-*TNANGNNAGGCTNNCCT*-3'), the so-called iron box, and the regulation of important metabolic genes devoted to iron assimilation is so intimate that we have previously exploited this feature to mine bacterial genomes and gain insights about their possible functions and molecular mechanisms (Barona-Gómez et al. 2006; Tierrafria et al. 2011; Cruz-Morales et al. 2017). This approach will be described in the proof-of-concept section of this chapter.

dFOs not only have important physiological roles but also morphological implications in *Streptomyces* and other myceliated *Actinobacteria*. Under laboratory conditions, *Amycolatopsis sp.* AA4 (originally confused with a *Streptomyces* species) produces a catechol-peptide siderophore called amychelin, which can sequester dFO-E produced by *Streptomyces coelicolor*. In turn, production of amychelin is arrested, and morphological changes take place through the action of the pleiotropic regulator *bldN* (Traxler et al. 2012; Lambert et al. 2014). In agreement with this, we have previously shown that mutation of *desE*, coding for the SBP acting as the main dFO receptor in *S. coelicolor* (Patel et al. 2010, Ronan et al. 2018), gives rise to a bald or *bld* mutant (Tierrafria et al. 2011). Given that DesE represents the only SBP capable of sensing dFO-E complexes (i.e., the cyclic form), it is tempting to speculate that this node, at the interface between the environment and the cell, integrates nutritional information with morphological development.

The implication of this possibility is that dFO-E may also be a morphogene and not only a metabolite useful to meet a nutritional requirement. Thus, DesE, encoded in the dFO or *des* BGC, which transports dFO-E plus all linear dFOs, as well as other hydroxamate siderophores not produced by *Streptomyces* (Barona-Gomez et al. 2006; Patel et al. 2010; Tierrafría et al. 2011; Arias et al. 2015; Ronan et al. 2018), may be facing an adaptive conflict. The proposed model shown in Fig. 10.2 reconciles this multifunctionality, believed to occur only in myceliated microorganisms, with the multispecific nature of SBPs, typically encountered in any microorganism evolving under iron limitation, such as that encountered in CCB. This model includes CchF, the SBP encoded in the coelichelin or *cch* gene cluster (Lautru et al. 2005), implicated in the specific transport of ferricoelichelin (Barona-Gomez et al. 2006; Patel et al. 2010), and CdtB, a BGC-independent transporter, capable of transporting dFO-B and ferricoelichelin (Bunet et al. 2006; Tierrafría et al. 2011).

Overwhelmed by the biological complexity related to the hydroxamate siderophores coelichelin and desferrioxamines, which must relate to the complexity of any microbial ecological system mediated by specialized metabolites (Pessotti et al. 2018), we have previously proposed that evolutionary analysis can bring about novel biological insights with functional and ecological implications (Cruz-Morales et al. 2017). In the following sections, we illustrate this idea using novel genomic data of *Actinobacteria* strains isolated from CCB.

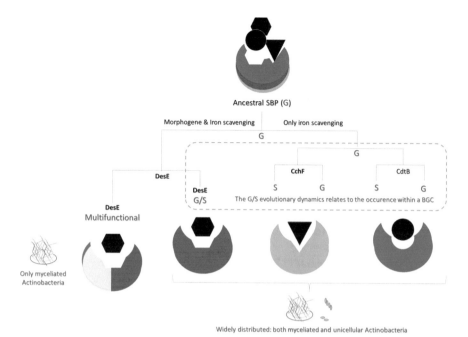

Fig. 10.2 Model for the evolution of siderophore-binding proteins (SBPs). The natural history of SBPs is shown as a simplified cartoon. Function (signaling and iron scavenging) and specificity (S specialist, G generalist) are shown at the nodes of the tree and at the tips. SBP functions and specificities are shown as cartoons under the tree. Black shapes represent diverse hydroxamate siderophores, while actinobacterial morphology (myceliated and unicellular associated) to SBPs is shown at the bottom. These proposed patterns, which are supported by phylogenetic analysis reported by Cruz-Morales et al. (2017), suggest that SBPs evolved from a generalist ancestor devoted to iron acquisition, which may seem functionally closer to CdtB. Such an ancestor must have been able to transport a wide range of siderophores, while unrelated to their biosynthesis. Divergence of SBPs occurred later with the appearance of DesE, which became fixed only in myceliated species, and of CchF and CdtB in a broader range of *Actinobacteria*, both myceliated and unicellular

An Evolutionary Model to Simplify Siderophore Ecological Studies

Our proposed model for the diversification of SBPs links dFO chemical diversity with the wealth of SBPs by taking into account the multifunctionality and multi-specificity of siderophores (Fig. 10.2). This model proposes that bacterial SBPs evolved from a generalist ancestor devoted to acquire ferric iron from the environment as ferrous iron became scarce. Once able to transport a wide range of sidero-phores, irrespective of their synthesis (which could have occurred abiotically or as a metabolic side product of primordial bacteria, e.g., citrate), this ancestral SBP would "functionally" resemble CdtB. This primordial SBP, as suggested by its modern form, must have been efficient in the transport of many hydroxamate siderophores, serving as a generalist, but also may have given place to both specialized transporters associated with specific BGCs, as well as stand-alone SBPs acting as generalists. The latter may explain the generalized ability of microorganisms to engage in "biopiracy" of xenosiderophores (Barona-Gómez et al. 2006), whereas the former relates to exquisite examples in which biosynthesis is coupled with efficient transport mechanisms via tight regulatory gene networks.

Thus, the evolution of hydroxamate siderophores can be used to analyze environmental genomic data from CCB as follows: (i) all hydroxamate SBP families include both BGC-associated and stand-alone SBPs; (ii) BGC-*desE* homologs are never associated with ATPase or permease components needed for a complete ABC transport system, which contrasts with stand-alone *desE* scenarios; this contrasting scenario involves the multifunctionality of DesE related to morphological development; and (iii) the BGC-*cdtB* and BGC-*cchF* homologs seem to be both recently recruited and evolved and may have a functional connection with the products of their cognate BGCs. Beyond the validity of these evolutionary hypotheses, which are hard to test (e.g., which SBP at the sequence level is closer to the ancestor of this large protein family?), the importance of these corollaries is that they provide a framework to analyze genomic data directly obtained from the environment, either as genome sequences of novel isolates, or even as metagenomes.

For instance, given the association between nutritional stress with developmental differentiation in myceliated *Actinobacteria*, and the exclusive occurrence of DesE homologs in these type of bacteria, it is tempting to speculate that the DesE receptors evolved from iron scavengers to morphogens, such as dFO-E. While conserving their ability to bind different hydroxamate siderophores, they started to play a regulatory role via an unknown mechanism, becoming a multifunctional protein. The latter scenario implies an adaptive conflict (Des Marais and Rausher 2008), as a transport mechanism that evolved for iron acquisition was co-opted for developmental signaling in microorganisms that undergo morphological differentiation. In this scenario, appearance of other co-occurring siderophores in myceliated *Actinobacteria* as previously noted (e.g., as coelichelin associated with its cognate specialized SBP CchF) may have occurred as an evolutionary mechanism to deal with this adaptive conflict. Indeed, it has been shown that, in *Streptomyces coelicolor,* coelichelin is the dominant siderophore produced for iron acquisition (Traxler et al. 2012). Thus, rather than contingency, as previously proposed (Challis and Hopwood 2003), we provide an evolutionary framework that takes into account adaptive conflicts, which is capable of reconciling evolution of multifunctionality and multispecificity, explaining the wealth of siderophore biosynthetic and transport systems present in *Actinobacteria*.

Proof-of-concept: Complex Community Interplay in CCB Made Simpler in the Light of Evolution

In our most recent study, we performed a phylogenomics analysis of hydroxamate siderophore biosynthesis and transport genes present in *Actinobacteria* (Cruz-Morales et al. 2017). For this purpose, we used a genomic database enriched with non-model microorganisms isolated from CCB. These habitats are extremely limited not only in phosphorus but also in iron (Souza et al. 2008) and provide a nearly ideal scenario to test our evolutionary model. The sequenced organisms were isolated from different ponds that compose the CCB hydrological system, as well as their terrestrial surroundings, as sampled between November 2012 and March 2016. Culture conditions included mainly those typically used for isolation of *Actinobacteria*, plus media limited in iron (Table 10.1; see also Cruz-Morales et al. 2017). The media used allows isolation of both myceliated and non-myceliated morphotypes. As iron is actually present in glassware, including beakers and flasks, its removal needs to be ensured via the chelation of ferric iron with either an appropriate resin during media preparation (and use of plasticware) or by the use of strong iron-chelators, typically 2,2′ bipyridyl, supplemented into the medium. The same principle of iron chelation is exploited during the so-called chrome azurol assay (*O*-CAS) (Schwyn and Neilands 1987; Louden et al. 2011), with the advantage of the formation of colored complexes that can be used to reveal the production of hydroxamate siderophores (e.g., Senges et al. 2018).

We took on the task of testing our evolutionary model using genomic data generated from strains isolated from the ponds referred to as Pozas Rojas, Pozas Azules, and Churince (Table 10.1). Isolation of the microbial diversity grown on a selective medium containing the iron-chelating compound 2,2′ bipyridyl (at a concentration of 200 uM) allowed us to obtain bacteria able to produce and/or transport siderophores. These bacteria were isolated, axenically cultivated, and taxonomically characterized after amplification and sequencing of their 16S ribosomal RNA gene. In parallel, we screened for the ability of these bacteria to produce siderophores using both the low-resolution *O*-CAS assay and high-resolution HPLC-MS metabolomic analysis. The latter allowed us to purify and characterize siderophores, including their chemical structure(s) based in previous reports (Fig. 10.3). Among the isolated microorganisms, some reported here for the first time, we focus our attention on three *Actinobacteria* strains whose genomes were sequenced and analyzed. These selected bacteria include two strains of the genus *Lentzea*, as well as one belonging to the genus *Kocuria*: (i) *Lentzea* sp. PRa11 and (ii) *Lentzea* sp. PRa55, which were isolated from the immediate surroundings of Pozas Rojas, and (iii) *Kocuria* sp. CH10, obtained from groundwater near Churince pond.

Table 10.1 Novel isolates from CCB obtained under iron-limited conditions

ID	Taxonomic filiation (16 s)	General morphology	Origin	Geographic isolation site	Isolation date
CH5002	*Streptomyces patensis*	White mycelium	Hydrological system	Churince	March 2016
CH5003	*Streptomyces* sp. *m5-1av*	White mycelium	Groundwater	Churince	March 2016
CH5005	*Nocardiopsis alkaliphilla*	Hard transparent spores	Hydrological system	Churince	March 2016
CH5007	*Nocardia seriolae*	Spores white rounded	Hydrological system	Churince	March 2016
CH5024	*Streptomyces viridodiastaticus*	White mycelium	Hydrological system	Churince	March 2016
PR5035	*Streptococcus* sp.	White and green mycelium	Hydrological system	Pozas Rojas	March 2016
CC51	*Streptomyces globisporus*	Light brown non-mycelium	Groundwater	Pozas Rojas	March 2016
PR5001	*Streptomyces SHXFF-2*	Black litmus mycelium	Hydrological system	Pozas Rojas	March 2016
PR5007	*Streptomyces pluripotens*	Red and hard colonies	Groundwater	Pozas Rojas	March 2016
CC036	*Streptomyces rochei*	Spores white rounded	Stromatolite	Poza Azul	March 2013
PR083	*Actinokineospora diospyrosa*	Orange, not mycelial	Groundwater	Pozas Rojas	October 2013
PRa55*	*Lentzea* sp.	White colonies of orange background removable mycelium	Surroundings of the hydrological system	Pozas Rojas	March 2013
PRa11*	*Lentzea jiangxiensis*	Spores white rounded	Surroundings of the hydrological system	Pozas Rojas	March 2014
CH10*	*Kocuria* sp. *G6*	Non-sporulated and not mycelial	Groundwater	Churince	October 2013

The table includes bacterial strains isolated in iron-limited media from samples obtained in CCB, including the ponds and pozas, their immediate surroundings, groundwater, and stromatolites. All these bacteria have the capacity to produce or transport siderophores. General strain identification was done using 16S ribosomal RNA sequences compared against the nonredundant database of GenBank; organisms marked with an asterisk (*) are taxa whose genomes (available upon request) were sequenced and analyzed in this chapter

On one hand, *Lentzea* is a relatively novel filamentous actinobacterium and represents a mesophilic genus from the family *Pseudonocardiaceae*. The genus *Lentzea* was proposed for aerobic actinobacteria that form abundant aerial hyphae that fragment into rod-shaped elements (Yassin et al. 1995; Li et al. 2018). It has also been reported as an efficient producer of peptides with antibiotic activity (Moniciardini et al. 2014; Zhao et al. 2018; Hussain et al. 2018), while three phenolated siderophores were previously reported during analysis of *Lentzea jiangxiensis*, isolated

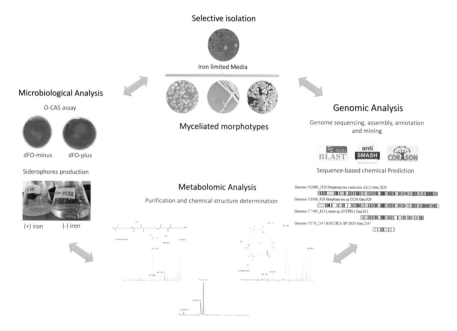

Fig. 10.3 Workflow and basic strategy for growth and isolation of siderophore-producer and transporter strains. Samples from groundwater, stromatolites, and hydrological systems from CCB were collected (see also Table 10.1) and used to selectively isolate *Actinobacteria* in iron-limited and SFM media. The ability of the isolates to produce siderophores was addressed using the *O*-CAS assay (Louden et al. 2011, Cruz-Morales et al. 2017). The presence of orange halos indicates the production of hydroxamate siderophores. To stimulate the production of siderophores in a fermentation broth and to obtain extracts for LC-MS analysis, cultures with and without 2,2′ bipyridyl were grown. In parallel, the genomic DNA of selected strains was extracted, assembled (Velveth), and annotated (RAST/antiSMASH). To further assess roles in iron metabolism, for both the predicted siderophore BGCs and SBPs, we identified the regulatory sequences or iron boxes using Artemis genome browser. Putative chemical products of BGCs were inferred based on sequence analyses

from red soils in China (Guo et al. 2015). On the other hand, the genus *Kocuria* (Stackebrandt et al. 1995; earlier classified together with the genus *Micrococcus*) are coccoid, aerobic, non-encapsulated, and non-endospore-forming bacteria, belonging to the order *Actinomycetales*. Species of the genus have been isolated from a broad range of sources (Kaur et al. 2011). A *Kocuria* sp. isolated from a saline desert has been reported to produce large amounts of indole-3-acetic acid, an auxin present in plant growth-promoting bacteria (Goswami et al. 2014; Gouda et al. 2018), as well as dFO-E directed by the *des* BGC (Boonlarppradab 2007; Caspi et al. 2010). This bacteria therefore seem not to be specific to the CCB, despite characteristic features that may explain their adaptation to this unique niche.

As depicted in Fig. 10.3, once bacteria are isolated and screened for a certain phenotype (in this case siderophore-mediated iron acquisition), their genomes are sequenced and analyzed. As the success of these analyses depends on the genomic diversity, de novo sequence assemblies are preferred over assemblies using reference genome sequences. The risk of missing important genes accompanies the latter strategy as previously demonstrated during sequencing and analysis of *Streptomyces lividans* (Cruz-Morales et al. 2013). More importantly, analysis of the genomic diversity may allow us to better understand how bacteria may interact but only after a high-quality genome annotation process is completed. In this context, there are as many approaches and tools for this purpose, as there are biological questions to be addressed: in this example, with an emphasis on iron metabolism, we exploit the predictive power of the iron boxes as a "beacon" or signal to confidently annotate gene function with regard to iron-related molecular processes. In turn, despite the overwhelming amount of data provided by genomic or metagenomic sequencing, this approach allows to efficiently direct the analysis increasing the likelihood of tackling complex biological and ecological questions. This process can be iteratively performed as novel biological and chemical data are incorporated.

After careful analysis of the obtained genomes of our selected *Lentzea* and *Kocuria* strains, we did actually find a series of patterns that help to illustrate the predictive power of our evolutionary model. It should be noticed that this analysis was undertaken assuming very conservative criteria in which only functional annotations of which we are fully confident were considered (Fig. 10.4). For this purpose, in addition to searching for iron boxes, we used our in-house platform, CORASON (*COR*e *A*nalysis of *S*yntenic *O*rthologs to prioritize *N*atural Product Biosynthetic Gene Cluster), to perform comparative phylogenomics centered at all the biosynthetic and transport genes that we could identify. Examples and complementary information related to this approach for other *Actinobacteria* of CCB (Cruz-Morales et al. 2017) and *Pseudomonas* plant endophytes (Gutierrez-Garcia et al. 2017) have been previously reported. It is worthwhile to emphasize the importance of sufficient data to capture the diversity of the system (which can be measured using rarefaction curves) as well as the structure of the ecological niche, if possible, both physically and genetically. Within this context, the full potential of phylogenomics and metagenomics is harnessed (Cibrian-Jaramillo and Barona-Gómez 2016), closing the gap between ecological and functional analyses.

The use of only two *Lentzea* genomes from the same site, yet not necessarily from the same ecological micro-niche, plus a *Kocuria* genome from a connected yet distant site (Pozas Rojas and Churince ponds, respectively) imposes many limitations to this analysis. Nevertheless, these data serve the purpose of this chapter, allowing us to gain interesting insights about the ecology of siderophore-mediated iron acquisition by *Actinobacteria* in CCB. As can be seen in Fig. 10.4, there are three fundamental observations, each of which were manually confirmed. *First,* strains PRa11 and PRa55, both belonging to the genus *Lentzea* and isolated from the same site (Pozas Rojas), have a different complement when it comes to the *des* and *cch* BGCs, including their cognate SBPs DesE and CchF. Strain PRa55 lacks a *des* BGC, responsible for the synthesis of dFOs. However, its genome encodes a stand-alone *desE* gene within an operon that includes the remaining ABC genes (permeases and ATPase), never present in the canonical actinobacterial *desEF-desABCD(G)* BGC. Moreover, the genome of PRa55 encodes a *cch* BGC that includes the *cchF*

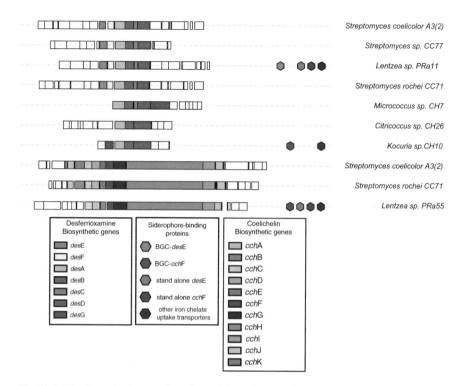

Fig. 10.4 The *des* and *cch* genomic regions of *Actinobacteria* from CCB. We used CORASON to compare the *des* and *cch* BGCs from *Streptomyces coelicolor* A3(2) with their equivalent loci found in the CCB strains. Homologous genes are displayed in the same color. As mentioned in the text, there are three main observations: (1) *Lentzea* sp. PRa11 and *Lentzea* sp. PRa55 carry different siderophore BGCs. While PRa11 has a complete *des* BGC, PRa55 has a *cch* BGC. (2) Even though *Kocuria* sp. CH10 possess a *des* BGC, it has a *cchF* gene (in blue) instead of a canonical *desE* gene (in green). (3) in the three strains analyzed, besides the BGC-related SBPs, we found stand-alone gene copies of SBPs (indicated as hexagons), including those for DesE, CchF, CdtB, and other iron chelate uptake transporters

gene. In contrast, strain PRa11 encodes for a complete *des* BGC, as well as an extra copy of a stand-alone *desE* gene, but not a *cch* BGC. *Second,* and quite remarkably, the *desE* homolog that is situated within the *des* BGC of the *Kocuria* CH10 strain, a non-myceliated actinobacteria, is closer at the sequence level to CchF than DesE. This came as a surprise, which could be confirmed in all other *Kocuria* genomes available to date, meaning that it was overlooked by our previous analyses rather than being an exception. Interestingly, in contrast with the canonical actinobacterial *des* BGC, the *Kocuria* hybrid *des-cchF* cluster actually includes the remaining ABC genes (Fig. 10.5). *Third,* as expected, all three strains have extra copies of several iron chelate uptake transporters, not only of the *cdtB* family, in association with the needed ABC genes, to make this generalist SBP functional.

The implication of these observations is that our model did actually fit with the reality encountered for these strains from CCB. The data support a notion of adap-

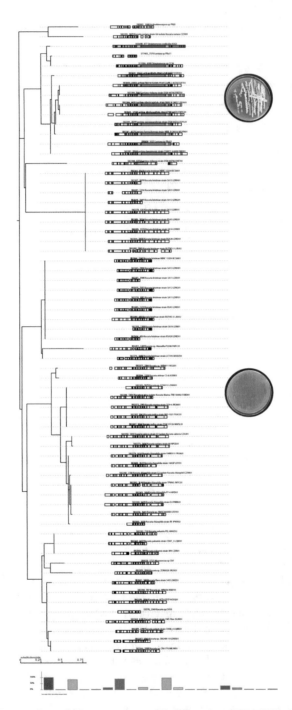

Fig. 10.5 Phylogenomic analysis centered at *cchF* SBPs using CORASON. CORASON is a visual tool that searches for gene clusters that are similar to a reference. The output of this

tive conflict encountered by DesE in terms of its multifunctionality, as suggested by the fact that DesE is replaced by CchF in non-myceliated bacteria, leading to the so-called hybrid *des-cchF* BGC. They also tell a story about how closely related species may interact via specialized metabolites. Both strains from the genus *Lentzea,* potentially co-existing in the same ecological niche, may interact via the exchange of siderophores, as both the biosynthetic machineries and transport capabilities of these strains are complementary. These observations, together with our previous analyses in CCB, pave the way to defining in more detail the role of iron in this unique community, where iron limitation seems to have prompted an explosion of all classes of SBPs, both generalist and specialist. It remains to be seen the role of other chemical types of siderophores in this model ecological system.

Final Remarks

The combined approach presented here to investigate the role of hydroxamate siderophores in CCB, which involves microbiology, genomics, chemistry, and evolutionary analysis, is paving the way to tackle complex ecological questions. CCB provides an optimal ecological system for this purpose, as it fulfills many of the criteria that have been recently discussed in the literature for this purpose (Pessotti et al. 2018). However, in light of our CCB results, there are many remaining questions to be addressed. For instance, it would be interesting, and feasible as judged from the current analytical state-of-the-art techniques, to systematically investigate in the laboratory interactions between *Actinobacteria* and other co-existing bacteria, from both distantly related taxa and well-defined common micro-niches. The latter is particularly overlooked due to the difficulty in defining the physical barriers of any given micro-niche. Also, now that target genes have been identified, it would be informative to investigate the ecological dynamics of key iron functional genes to learn, at the population level, their role in CCB. This could be complemented by assessment in situ of the concentration of hydroxamate siderophores in various ponds and other sites, as it has previously been done in oceanic environments to correlate bacterial interactions with iron limitation (Boiteau et al. 2016).

Fig. 10.5 (continued) comparative genomic analysis, which includes publically available *Lentzea* and *Kocuria* genomes as well as our own CCB actinobacterial genomes, expands the observations of Fig. 10.4. Aside from searching for similar clusters, CORASON also constructs a phylogenetic tree, concatenating the query sequence and the core genes of all clusters displayed. For this analysis, we used the *cchF* sequence as query, and similar colors indicate homologous genes. The top sub-clade shows the typical *cch* BGC mostly found in myceliated organisms, such as in strain PRa55 (top right-hand picture). The bottom clade shows all the hybrid *des-cchF* BGCs found in *Kocuria* strains that are not myceliated, including strain CH10 (bottom right-hand picture). Please note that *des* genes are not colored, because the *cchF* and the *Streptomyces coelicolor* A3(2) coelichelin BGC were used by CORASON as references. Only similar genetic contexts found in all detected clusters are colored. In between these two clades, stand-alone copies of *cchF* in *Kocuria* that were also detected by this analysis are shown

Acknowledgments The work by our laboratory in the CCB was made possible, thanks to the support by Valeria Souza and Gabriela Olmedo and their teams and funding to FBG from Conacyt grants (Nos. 179290 & 285746). The authors would like to thank the many members of the Evolution of Metabolic Diversity Laboratory that took part in expeditions to CCB during 2012–2016 and subsequent experimental and bioinformatics work. We would like to specially thank Alejandra Castañeda, Pablo Cruz, Milan Janda, Cuauhtémoc Licona, Paulina Mejía, Sandra Pérez, Hugo Ramírez, José-Luis Steffani, Nelly Selem, Karina Gutiérrez, and Mariana Vallejo, without whom this work could not have been possible.

References

Ahmed E, Holmström SJM (2014) Siderophores in environmental research: roles and applications. Microb Biotechnol 7:196–208

Arias AA, Lambert S, Martinet L et al (2015) Growth of desferrioxamine deficient *Streptomyces* mutants through xenosiderophore piracy of airborne fungal contaminations. FEMS Microbiol Ecol 91:fiv080

Barona-Gómez F, Wong U, Giannakopulos AE et al (2004) Identification of a Cluster of Genes that Directs Desferrioxamine Biosynthesis in *Streptomyces coelicolor* M145. J Am Chem Soc 126:16282–16283

Barona-Gómez F, Lautru S, Francou FX, Leblond P, Pernodet JL, Challis GL et al (2006) Multiple biosynthetic and uptake systems mediate siderophore-dependent iron acquisition in *Streptomyces coelicolor* A3(2) and *Streptomyces ambofaciens* ATCC 23877. Microbiology 152:3355–3366

Boiteau R, Mende D, Hawco N et al (2016) Siderophore-based adaptations to iron scarcity. PNAS 113:14237–14242

Boonlarppradab C (2007) Investigation of the potential anticancer and antifungal active secondary metabolites from marine natural products. Peer reviewed/Thesis/dissertation, UC San Diego

Bruns H, Crüsemann M, Letzel A et al (2018) Function-related replacement of bacterial siderophore pathways. ISME J 12:320–329

Bunet R, Brock A, Rexer HU et al (2006) Identification of genes involved in siderophore transport in Streptomyces coelicolor A3(2). FEMS Microbiol Lett 262:57–64

Caspi A, Hariri AR, Holmes A et al (2010) Genetic sensitivity to the environment: the case of the serotonin transporter gene and its implications for studying complex diseases and traits. Am J Psychiatry 167:509–527

Challis G, Hopwood D (2003) Synergy and contingency as driving forces for the evolution of multiple secondary metabolite production by *Streptomyces* species. PNAS 100(Suppl 2):14555–14561

Cibrián-Jaramillo A, Barona-Gómez F (2016) Increasing metagenomic resolution of microbiome interactions through functional phylogenomics and bacterial sub-communities. Front Genet 10(7):4

Codd R, Richardson-Sanchez T, Telfer TJ et al (2018) Advances in the chemical biology of Desferrioxamine B. ACS Chem Biol 13:11–25

Cornelis P, Andrews SC (2010) Iron uptake and homeostasis in microorganisms. Caister Academic Press, Norfolk

Cruz-Morales P, Vijgenboom E, Iruegas-Bocardo F et al (2013) The genome sequence of S*treptomyces livi*dans 66 reveals a novel tRNA-dependent peptide biosynthetic system within a metal-related genomic island. Genome Biol Evol 5:1165–1175

Cruz-Morales P, Ramos-Aboites HE, Licona-Cassani C et al (2017) Actinobacteria phylogenomics, selective isolation from an iron oligotrophic environment and siderophore functional characterization, unveil new desferrioxamine traits. FEMS Microbiol Ecol 93(9):fix086. https://doi. org/10.1093/femsec/fix086

D'Onofrio A, Crawford JM, Stewart EJ et al (2010) Siderophores from neighboring organisms promote the growth of uncultured bacteria. Chem Biol 17:254–264

Des Marais DL, Rausher MD (2008) Evidence for escape from adaptive conflict. Nature 454:762–765

Essen SA, Johnsson A, Bylund D et al (2007) Siderophore production by *Pseudomonas stutzeri* under aerobic and anaerobic conditions. Appl Environ Microbiol 73:5857–5864

Fardeau S, Mullié C, Dassonville-Klimpt A et al (2011) Bacterial iron uptake: a promising solution against multidrug resistant bacteria. In: Méndez-Vilas A (ed) Science against microbial pathogens: Communicating current research and technological advances. Formatex, Badajoz, pp 695–705

Galet J, Deveau A, Hotel L et al (2015) Pseudomonas fluorescens pirates both ferrioxamine and ferricoelichelin siderophores from Streptomyces ambofaciens. Appl Environ Microbiol 81:3132–3141

Goswami D, Pithwa S, Dhandhukia P et al (2014) Delineating Kocuria turfanensis 2M4 as a credible PGPR: a novel IAA-producing bacterium isolated from saline desert. J Plant Interact 9:566–576

Gouda S, Kerry R, Das G et al (2018) Revitalization of plant growth promoting rhizobacteria for sustainable development in agriculture. Microbiol Res 206:131–140

Gubbens J, Wu C, Zhu H et al (2017) intertwined precursor supply during biosynthesis of the catecholate-hydroxamate siderophores qinichelins in Streptomyces sp. MBT76. ACS Chem Biol 12:2756–2766

Guo X, Liu N, Li X et al (2015) Red soils harbor diverse culturable actinomycetes that are promising sources of novel secondary metabolites. Appl Environ Microbiol 81:3086–3103

Gutiérrez-García K, Neira-González A, Pérez-Gutiérrez R et al (2017) Phylogenomics of 2,4-Diacetylphloroglucinol-Producing Pseudomonas and Novel Antiglycation Endophytes from Piper auritum. J Nat Prod 80:1955–1963

Hider RC, Kong X (2010) Chemistry and biology of siderophores. Nat Prod Rep 27:637–657

Holden VI, Bachman MA (2015) Diverging roles of bacterial siderophores during infection. Metallomics 7:986–995

Hussain A, Rather M, Dar M et al (2018) *Streptomyces puniceus* strain AS13., Production, characterization and evaluation of bioactive metabolites: A new face of dinactin as an antitumor antibiotic. Microbiol Res 207:196–202

Kadi N, Oves-Costales D, Barona-Gómez F et al (2007) A new family of ATP-dependent oligomerization – macrocyclization biocatalysts. Nat Chem Biol 3:652–656

Kaur C, Kaur I, Raichand R et al (2011) Description of a novel actinobacterium *Kocuria assamensis* sp. nov., isolated from a water sample collected from the river Brahmaputra, Assam, India. Antonie Van Leeuwenhoek 99:721–726

Lambert S, Traxler MF, Craig M et al (2014) Altered desferrioxamine-mediated iron utilization is a common trait of bald mutants of *Streptomyces coelicolor*. Metallomics 6(8):1390–1399

Lautru S, Deeth RJ, Bailey LM et al (2005) Discovery of a new peptide natural product by *Streptomyces coelicolor* genome mining. Nat Chem Biol 1:265–269

Li D, Zheng W, Zhao J et al (2018) *Lentzea soli* sp. nov., an actinomycete isolated from soil. Int J Syst Evol Microbiol 68:1496–1501

Louden B, Haarmann D, Lynne AM (2011) Use of blue agar CAS assay for siderophore detection. J Microbiol Biol Educ 12:51–53

Messenger A, Barclay R (1983) Bacteria, iron and pathogenicity. Biochem Educ 11:54–63

Miethke M, Marahiel MA (2007) Siderophore-based iron acquisition and pathogen control. Microbiol Mol Biol Rev 71:413–451

Monciardinni P, Iorio M, Maffioli S et al (2014) Discovering new bioactive molecules from microbial sources. Microb Biotechnol 7:209–220

Patel P, Song L, Challis GL (2010) Distinct extracytoplasmic siderophore binding proteins recognize ferrioxamines and ferricoelichelin in *Streptomyces coelicolor* A3(2). Biochemistry
  49:8033–8042
Pessotti R, Hansen BL, Traxler MF (2018) In search of model ecological systems for understanding specialized metabolism. mSystems 3(2):e00175–e00117
Roberts AA, Schultz AW, Kersten RD et al (2012) Iron acquisition in the marine actinomycete genus *Salinispora* is controlled by the desferrioxamine family of siderophores. FEMS
  Microbiol Lett 335:95–103
Ronan J, Kadi N, McMahon S et al (2018) Desferrioxamine biosynthesis: diverse hydroxamate
  assembly by substrate-tolerant acyl transferase DesC. Philos Trans R Soc Lond B Biol Sci
  373:20170068
Saha M, Sarkar S, Sarkar B et al (2016) Microbial siderophores and their potential applications: a
  review. Environ Sci Poll Res 23:3984–3999
Schwyn B, Neilands B (1987) Universal chemical assay for the detection and determination of
  siderophores. Anal Biochem 160:47–56
Senges C, Al-Dilaimi A, Marchbank D et al (2018) The secreted metabolome of *Streptomyces
  chartreusis* and implications for bacterial chemistry. PNAS 115:2490–2495
Smits TH, Duffy B (2011) Genomics of iron acquisition in the plant pathogen *Erwinia amylovora*
  insights in the biosynthetic pathway of the siderophore desferrioxamine E. Arch Microbiol
  193:693–699
Souza V, Eguiarte L, Siefert J et al (2008) Microbial endemism: does phosphorus limitation
  enhance speciation? Nat Rev Microbiol 6(2008):559–564
Stackebrandt E, Koch C, Gvozdiak O et al (1995) Taxonomic dissection of the genus Micrococcus:
  Kocuria gen. nov., Nesterenkonia gen. nov., Kytococcus gen. nov., Dermacoccus gen. nov., and
  Micrococcus Cohn 1872 gen. emend. Int J Syst Bacteriol 45:682–692
Tierrafría VH, Ramos-Aboites HE, Gosset G et al (2011) Disruption of the siderophore-binding
  desE receptor gene in *Streptomyces coelicolor* A3(2) results in impaired growth in spite of
  multiple iron–siderophore transport systems. Microb Biotechnol 4:275–285
Traxler M, Kolter R (2015) Natural products in soil microbe interactions and evolution. Nat Prod
  Rep 32:956–970
Traxler MF, Seyedsayamdost MR, Clardy J et al (2012) Interspecies modulation of bacterial devel-
  opment through iron competition and siderophore piracy. Mol Microbiol 86:628–644
Traxler MF, Watrous JD, Alexandrov T et al (2013) Interspecies interactions stimulate diversification of the *Streptomyces coelicolor* secreted metabolome. MBio 20:4
Tunca S, Barreiro C, Sola-Landa A et al (2007) Transcriptional regulation of the desferrioxamine
  gene cluster of *Streptomyces coelicolor* is mediated by binding of DmdR1 to an iron box in the
  promoter of the desA gene. FEBS J 274:1110–1122
Tunca S, Barreiro C, Coque JJ et al (2009) Two overlapping antiparallel genes encoding the iron
  regulator DmdR1 and the Adm proteins control siderophore and antibiotic biosynthesis in
  Streptomyces coelicolor A3(2). FEBS J 276: 4814–4827
Wang W, Qiu Z, Tan H et al (2014) Siderophore production by actinobacteria. Biometals
  27:623–631
Yamanaka K, Oikawa H, Ogawa HO et al (2005) Desferrioxamine E produced by *Streptomyces
  griseus* stimulates growth and development of *Streptomyces tanashiensis*. Microbiology 151(Pt
  9):2899–2905
Yassin AF, Rainey FA, Brzezinka H et al (1995) Lentzea gen. nov., a new genus of the order
  Actinomycetales. Int J Syst Bacteriol 45:357–363
Zhao P, Xue Y, Gao W et al (2018) Actinobacteria-Derived peptide antibiotics since 2000. Peptides
  103:48–59

Chapter 11
Animal-Mediated Nutrient Cycling in Aquatic Ecosystems of the Cuatro Ciénegas Basin

Eric K. Moody, Evan W. Carson, Jessica R. Corman, and Hector Espinosa-Pérez

Contents

Abstract Consumers can alter nutrient cycling and create biogeochemical hotspots by excreting nutrients such as nitrogen and phosphorus at varying rates and stoichiometric ratios. Variation in these rates and ratios is in turn driven by traits such as body size, diet, and nutrient assimilation efficiency. These traits often vary considerably among and within species. As aquatic ecosystems in the Cuatro Ciénegas Basin are nutrient-poor, animals may play a large role in recycling key nutrients such as nitrogen (N) and phosphorus (P). We investigated this question by studying nutrient excretion rates of snails (*Mexithauma quadripaludium* and *Nymphophilus minckleyi*), a poeciliid fish (*Gambusia marshi*), and two cyprinodontid fishes (*Cyprinodon atrorus* and *C. bifasciatus*) over a variety of environments. Snails were collected from oncolites in the Río Mesquites, *C. atrorus* were measured in Laguna Intermedia, *C. bifasciatus* were measured in Poza La Becerra, and *G. marshi* were

E. K. Moody (✉)
Iowa State University, Department of Ecology, Evolution, and Organismal Biology, Ames, IA, USA

E. W. Carson
U.S. Fish and Wildlife Service, Bay-Delta Fish and Wildlife Office, Sacramento, CA, USA

J. R. Corman
University of Nebraska, School of Natural Resources, Lincoln, NE, USA

H. Espinosa-Pérez
La Universidad Autónoma de Mexico, Colección Nacional de Peces, Mexico D.F., Mexico

© Springer International Publishing AG, part of Springer Nature 2018
F. García-Oliva et al. (eds.), *Ecosystem Ecology and Geochemistry of Cuatro Cienegas*, Cuatro Ciénegas Basin: An Endangered Hyperdiverse Oasis,
https://doi.org/10.1007/978-3-319-95855-2_11

measured at nine sites around the basin as well as in laboratory experiments with fish from Poza La Becerra and Los Hundidos. We found substantial variation in nutrient excretion rates both among and within species and suggest that differences in diet and feeding rate could explain much of this variation. We also calculated the potential contribution of recycled P to supporting oncolite production and found that snails are likely not a major source of P to primary production in oncolites. Finally, we discuss how changes in these environments may lead to shifts in consumer-mediated nutrient recycling.

Keywords Aquatic ecosystem · Fishes · Phosphorus · Nutrient cycling · Snail

Introduction

Animals can play a large role in ecosystem-scale nutrient cycling by moving nutrients across the landscape and transforming biologically important nutrients into forms that are readily used by primary producers and heterotrophic microbes (Vanni 2002; Doughty et al. 2016). In aquatic ecosystems, the excretion of nitrogen (N) and phosphorus (P) in dissolved forms such as ammonia (NH_3) and soluble reactive phosphorus (SRP) can generate biogeochemical hotspots, reduce the severity of nutrient limitation, and even cause shifts from N- to P-limited primary production (Elser et al. 1988; McIntyre et al. 2008; Atkinson et al. 2013). Among ecosystems, the animals that have the largest influence on nutrient cycling can vary considerably; aquatic insects, crustaceans, mussels, tadpoles, fishes, and large mammals have been identified as having an important role in the nutrient cycles of aquatic ecosystems around the globe (e.g., Elser et al. 1988; Small et al. 2011; Atkinson et al. 2013; Whiles et al. 2013; Subalusky et al. 2017). As a result, it is important to understand what traits of organisms influence their importance to nutrient dynamics at the ecosystem scale. Considerable work on this subject has revealed that a combination of biomass, diet, mobility, and the fate of nutrients sequestered in an animal's body shape which species may contribute disproportionately to ecosystem function (McIntyre et al. 2008; Small et al. 2011; Vanni et al. 2013). Therefore, to identify which animals could play an important role in the function of aquatic ecosystems in Cuatro Ciénegas, we must examine which species exhibit unique combinations of these traits.

Aquatic ecosystems in the Cuatro Ciénegas Basin (CCB) are extremely nutrient-poor (Minckley and Cole 1968); thus, animals could play an especially important role in the nutrient cycles of these systems. As the waters of Cuatro Ciénegas are especially low in P, primary production in both the water column and on microbial oncolites is frequently P-limited (Elser et al. 2005a; Corman et al. 2016; Lee et al. 2017). Therefore, animals which transport and release P at high rates or in microhabitats where P is especially limiting could serve as keystone nutrient recyclers. While some large-bodied fishes such as the Cuatro Ciénegas cichlid (*Herichthys minckleyi*), largemouth bass (*Micropterus salmoides*), flathead catfish (*Pylodictis olivaris*), and headwater catfish (*Ictalurus lupus*) are distributed throughout much of CCB, with the exception of *H. minckleyi*, these tend to occur at low biomass

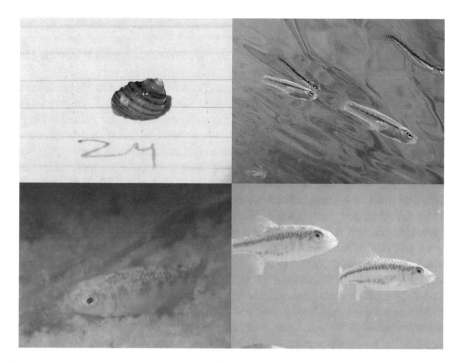

Fig. 11.1 The four focal animal species studied in this chapter. Clockwise from top left: *Mexithauma quadripaludium* (photo: Jessica Corman), *Gambusia marshi* (photo: Eric Moody), *Cyprinodon bifasciatus* (photo: Eric Moody), and *Cyprinodon atrorus* (photo: Eric Moody)

(Minckley 1969). In contrast, small-bodied fishes such as the robust gambusia (*Gambusia marshi*), Cuatro Ciénegas pupfish (*Cyprinodon bifasciatus*), and bolsón pupfish (*Cyprinodon atrorus*) can be quite abundant in appropriate habitats (Minckley 1969). While aquatic insects are not particularly abundant in most habitats in the basin, snails are also locally abundant in sediments and especially on oncolites (Hershler 1984; Dinger et al. 2005; Elser et al. 2005b). As a result, we will focus on nutrient recycling by abundant small-bodied fishes and snails in this chapter (Fig. 11.1), but we will discuss the potential importance of other animals in our concluding remarks.

Snails

Aquatic snails have drawn much attention in Cuatro Ciénegas. D. W. Taylor suggests that "the most spectacularly endemic fauna of freshwater snails known in the Western Hemisphere [is found in Cuatro Ciénegas]." In a paleoecological context, these organisms are a potentially paradoxical example of a grazer on stromatolites (Garcia-Pichel et al. 2004, Elser et al. 2005b), as one, albeit contested, theory for the decline of stromatolites after the Precambrian is metazoan grazing (e.g., Garrett 1970; Riding 2000). In any case, snails are abundant in the basin, and their feeding on oncoid microbialites may serve as an important mechanism for P recycling in these ecosystems.

In September 2010, 30 snails (*Mexithauma quadripaludium* and *Nymphophilus* sp.) were collected from oncoid microbialites in Río Mesquites, MX, and placed in petri dishes with filtered Río Mesquites water to determine nutrient excretion rates. Containers filled with filtered Río Mesquites water, but containing no animals, were used to correct for potential nutrient leaching or contamination. In 5 out of 30 containers, P excretion was not detected; these samples were removed from molar ratio calculations. Mean shell size (and standard deviation) was 4.70 (0.94) mm. In February 2011, the same experiment was repeated, but only *M. quadripaludium* were collected, and three animals were incubated in each container to achieve detectable nutrient concentrations. For some (8 out of 20) containers, NH_4^+ levels were still not above container blanks. For calculations, NH_4^+ concentrations for these samples were corrected to zero. SRP was not detected in one container; this sample was removed for molar ratio calculations. Snail shell size ranged from 1.70 to 4.86 mm with a mean (and standard deviation) of 3.15 (0.55) mm. There appears to be no allometric relationship between snail shell size (calculated by the sum of the three snails in the container) and excretion rates (although not presented, an allometric relationship was not detected in the September 2010 samples either).

The molar N:P ratio of snail excretion varied nearly two orders of magnitude, from 5 to 414. The mean N:P ratios were 79 in 2010 and 74 in 2011, which fall within the range of previously reported invertebrate nutrient excretion ratios (e.g., Devine and Vanni 2002). Mean nitrogen excretion rates (as the sum of NH_4^+-N and NO_3^--N) were 0.015 (0.006) µmoles N snail^{-1} h^{-1} in 2010 and 0.065 (0.029) µmoles N snail^{-1} h^{-1} in 2011. Mean phosphorus excretion rates were 0.004 (0.004) µmoles P snail^{-1} h^{-1} in 2010 and 0.001 (0.002) µmoles P snail^{-1} h^{-1}. There were no significant differences detected between N or P excretion between snail species in 2010.

Some rough calculations can be used to determine if snails play an important role in meeting autotrophic N and P demand. N and P demands have not been tested directly, so we first estimated it: Rates of primary production in an oncoid range from 5 mg O_2 L^{-1} h^{-1} in large (8–10 cm) oncoids (Elser et al. 2005a) to 0.98 mg O_2 L^{-1} h^{-1} in smaller (16.7 cm^3 or 2.31 cm) oncoids (Corman et al. 2016). Converted to area, this gives a primary production rate between 636.9 and 2364 mg O_2 L^{-1} h^{-1} m^{-2}. Assuming the higher limit of this range and that all of O_2 is converted to biomass with a C:P ratio of 500, there is a P demand of up to 0.296 mmol P day^{-1} m^{-2} in Río Mesquites. An average of 12 snails are found on an oncoid with a diameter of 8–10 cm (Elser et al. 2005a). By converting this to area, we can assume there are 1,528 snails per m^2 (diameter = 10 cm, area of a circle with a diameter of 10 cm = 78.5 cm^2). If each snail recycles 0.001 µmoles P h^{-1}, snails may recycle 1.53 µmoles P day^{-1} m^{-2}. Given the P demand calculated above, snails generate only 0.5% of P needed to support primary production in oncoids.

Fishes

Gambusia marshi

Within the Cuatro Ciénegas Basin, *Gambusia marshi* is almost certainly the most widespread and abundant species of fish (Minckley 1969). Beyond its abundance in a variety of habitats, *G. marshi* may also be important as a vector of nutrients from the terrestrial environment because a large proportion of their diet consists of terrestrial invertebrates (Meffe 1985; Hernández et al. 2017). While considerable work has examined the transfer of aquatic energy and nutrients to terrestrial systems, the transport of riparian nutrients into aquatic nutrient cycles via consumer nutrient excretion remains an under investigated question (Atkinson et al. 2017). Finally, this species is of theoretical interest because of its high degree of phenotypic variability among environments (Minckley 1962, 1969; Moody and Lozano-Vilano 2018), which could lead to variation in nutrient excretion rates among populations.

We measured the excretion rates and N:P ratios of adult female *G. marshi* from nine sites around the basin to examine variation in the field. This sampling was conducted at six spring sites (Anteojo, Escobedo, La Becerra, Las Teclitas, Mojarral Este, and Santa Tecla) as well as three runoff-fed wetlands (Laguna Intermedia, Los Gatos, and Los Hundidos) during May 2013. We found that for a given body mass, fish from springs excreted N at a lower rate and P at a higher rate (Moody et al. 2018; Fig. 11.2). As a result, spring fish excretion N:P was significantly lower than wetland fish excretion N:P (Moody et al. 2018). Modeling combined with a series of laboratory experiments described in Moody et al. (2018) suggest that this variation is driven by higher per capita consumption rates in springs, a hypothesis which is supported by the fact that *G. marshi* in Laguna Intermedia had significantly lower gut fullness relative to fish from the Churince spring (Hernández et al. 2017).

Fig. 11.2 Per capita *Gambusia marshi* excretion rates, excretion N:P, and water TDN:TDP of six springs and three wetlands sampled in Cuatro Ciénegas. Error bars represent standard error

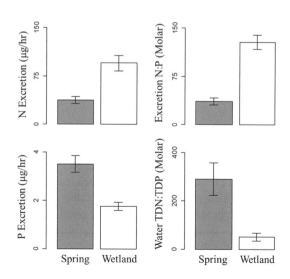

Moody et al. (2018) speculated that this variation may in turn be driven by reduced predation pressure in wetland environments where predatory largemouth bass and headwater catfish do not occur.

Having found that female *G. marshi* excrete nutrients at a lower N:P ratio in springs, we can examine whether this potentially impacts nutrient cycling by investigating the relationship between fish excretion N:P and water column N:P. If the excretion of these fishes alters nutrient limitation, we would expect that sites where *G. marshi* excretion N:P is higher to have a higher water column N:P as well. While we found no significant correlation between average excretion N:P and water column N:P measured in May 2013 ($r = -0.35$, $p = 0.359$), water column N:P was significantly higher in springs where fish excreted nutrients at a lower N:P ratio on average (Fig. 11.2). These results suggest that through the consumption of terrestrial arthropods, *G. marshi* could be an important source of P to P-limited spring systems, while their functional role is reduced in more P-rich wetland systems.

In addition to these field measurements, we also conducted laboratory experiments to investigate differences in the excretion rates of males and females. Wild fish were collected from La Becerra and Los Hundidos, and their F1 offspring were raised in the laboratory at both 25 °C and 32 °C. We measured the excretion rates of these fish upon reaching sexual maturity. Methods used to house fish and measure excretion rates are the same as those described by Moody et al. (2018). We measured excretion rates 3 h after feeding fish a uniform flake diet, as this was determined to capture the peak of post-ingestion excretion. We found that after correcting for body mass, N excretion rates did not vary between males and females ($F_{1,44} = 0.001$, $p = 0.977$), but P excretion rates were significantly higher in females ($F_{1,44} = 17.611$, $p < 0.001$). This difference likely reflects the fact that female *Gambusia* continue growing as adults, while males typically stop growing upon reaching maturity and thus consume food at lower rates than females. While our field measurements of *G. marshi* excretion rates primarily consisted of female fish, we also measured excretion rates of 4 males and 20 females from Lagunita, a shallow pond that was experimentally fertilized in 2012. Consistent with the laboratory experiments, we found that mass-specific P excretion rates were lower in males at this site. However, these data are limited in sample size and also confounded by the experimental fertilization; thus, they should be interpreted with caution.

Cyprinodon

Aside from *G. marshi*, the most abundant fishes in CCB are pupfish in the genus *Cyprinodon* (Minckley 1969). Two species in the genus are present in the basin: *C. atrorus*, which typically inhabits evaporative lagoons and wetlands, and *C. bifasciatus*, which typically inhabits spring pools and spring-fed river systems (Carson et al. 2008). While these species are generally allopatric, they have historically and continue to hybridize in intermediate habitats (Carson and Dowling 2006; Tobler and Carson 2010). Reciprocal transplants revealed that *C. atrorus* can survive in both spring and evaporative environments, suggesting that their absence from springs is due to greater predator susceptibility and/or competitive exclusion in these environments (Carson et al. 2008). Indeed, these pupfishes are ecologically similar in that they both primarily feed on benthic crustaceans and insects (Hernández et al. 2017). As a result, they are not a vector for terrestrial nutrients to enter aquatic ecosystems as are *G. marshi*, but they may still play an important role in recycling nutrients within the aquatic system.

We measured excretion rates of male and females of both species from the respective habitats where they are each most abundant. *C. atrorus* were measured in an evaporative lagoon, Laguna Intermedia, while *C. bifasciatus* were measured in a spring pool, Poza La Becerra. Methods used to measure excretion rates followed those for *G. marshi*. In total we measured excretion rates of 3 males and 4 females of *C. atrorus* and 11 males and 13 females of *C. bifasciatus*. These measurements were conducted in June 2012 and October 2011 for *C. atrorus* and *C. bifasciatus*, respectively.

We found that controlling for body size, variation in N excretion rates was determined by an interacting effect of sex and species. Females of *C. bifasciatus* excreted N at higher rates than males, but males of *C. atrorus* excreted N at higher rates than females. N excretion rates of *C. bifasciatus* were notably higher than for *C. atrorus* even after correcting for the fact that *C. bifasciatus* were larger. P excretion rates were also higher in *C. bifasciatus*, but we found no effect of sex on P excretion rates. As a result, N:P excreted did not vary between species or sexes. Excretion N:P was notably higher in both species of *Cyprinodon* than in any populations of *Gambusia marshi* (Fig. 11.3).

Conclusions

We found marked variation both among and within species of consumers in their rates of nutrient excretion as well as the N:P ratios of nutrients excreted. This variation may be driven by differences in diet, consumption rates, body stoichiometry, and other metabolic processes among habitats and species (Sterner 1990; Elser and Urabe 1999; Moody et al. 2015; Vanni and McIntyre 2016). Here, we will synthesize patterns in the data both within and among species and discuss their potential role in the nutrient cycles in CCB. Finally, we will conclude by discussing how changes to these systems may alter consumer-driven nutrient recycling across the basin.

On average, *G. marshi* excretion was relatively more P-rich (mean N:P = 25) than that of *C. atrorus* (mean N:P = 75), *C. bifasciatus* (mean N:P = 100), and snails (mean N:P = 74) (Fig. 11.3). However, there was considerable variation within species. Surprisingly, consistent patterns did not emerge in explaining this variation among species. As excretion rates depend on metabolic rate, they are expected to increase with body size with an approximate ¾ power scaling exponent (Allgeier et al. 2015; Vanni and McIntyre 2016). We found this was roughly true in *G. marshi*,

Fig. 11.3 Average per capita excretion rates and N:P ratios of snails (*Mexithauma quadripaludium*), pupfishes (*Cyprinodon atrorus* and *C. bifasciatus*), and gambusia (*Gambusia marshi*) from the Cuatro Ciénegas basin of Coahuila, Mexico. Error bars represent standard deviation

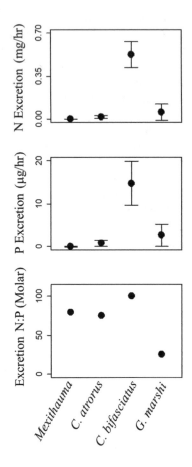

but not in the other species studied. Snail excretion rates did not vary with body size, while *Cyprinodon* excretion rates either did not vary or decreased with body size. In *Cyprinodon*, these effects are likely attributable to differences between sexes. N excretion rates decreased with size in *C. atrorus*, a species in which the females we measured were also larger than males. In *C. bifasciatus*, males were larger than females, and N excretion rate did not vary with body size. These results suggest that the effects of sex on nutrient excretion rates may not be consistent among species, but this is a topic worthy of further study.

Also inconsistent were the effects of habitat type on excretion rates and ratios among groups of fish. In *G. marshi*, fish from wetlands excreted N at a higher rate and P at a lower rate, leading to decreased excretion N:P relative to fish from springs (Fig. 11.2). In contrast, *C. atrorus* from wetlands excreted both N and P at lower rates relative to *C. bifasciatus* from springs. Further, mean *C. atrorus* excretion N:P was also lower than that of *C. bifasciatus*. These differences likely result from the different dietary habits of these two groups. *Gambusia* feed primarily on terrestrial material that falls onto the water surface, while *Cyprinodon* feed more heavily on benthic invertebrates, primarily chironomid larvae, as well as some zooplankton and plant material (Arnold 1972; Hernández et al. 2017). As a result, the diet of *G. marshi* likely varies less among habitats than that of *Cyprinodon*. Indeed, in Churince *C. bifasciatus* feed more heavily on N-rich copepods than *C. atrorus* (Hernández et al. 2017), which may contribute to their higher excretion N:P. Excretion N:P in fishes can vary as a function of both diet and consumption rate (Moody et al. 2015); thus, variation in consumption has been hypothesized to explain variation in excretion N:P within *G. marshi* (Moody et al. 2018). As these are the first published excretion rate measurements of any *Cyprinodon* pupfish, we have much to learn about what drives variation within and among species of these fascinating fishes.

Finally, we have made some very rough efforts to extrapolate our individual excretion rates to the ecosystem scale for one site, Río Mesquites, to estimate the relative importance of these species to P cycling in this system. Based on organismal density, we estimated that *M. quadripaludium* excrete 1.53 μmoles P day^{-1} m^{-2}, *G. marshi* excrete 25.86 μmoles P day^{-1} m^{-2}, and *C. bifasciatus* excrete 67.62 μmoles P day^{-1} m^{-2} (Fig. 11.4). Unfortunately, we lack *G. marshi* and *C. bifasciatus* density data from this system and thus used median values for closely related species (Kodric-Brown 1988; Araújo et al. 2014). We exclude *C. atrorus* from this analysis as they do not typically occur in Río Mesquites. These results suggest that *C. bifasciatus* may be particularly important P recyclers in springs due to their high abundance and high P excretion rates. Together, recycling by these organisms supplies approximately one third of estimated P demand in Río Mesquites (Fig. 11.4). While these areal calculations are very rough approximations that should only be used to get a relative estimate of the magnitude of areal excretion relative to demand, they indicate that these animals do play an important role in the P cycles of these spring systems. In particular, the abundant small-bodied fishes can serve as important vectors of terrestrial P as well as recyclers of P within the system.

The important role of animals in P cycling in the unique, P-poor ecosystems of Cuatro Ciénegas is especially significant considering the threats these species and systems face (Minckley 1992; Souza et al. 2006). Of these, *C. bifasciatus* is particularly threatened by groundwater extraction as their survival in environmentally

Fig. 11.4 The relative contribution of P recycling by *Cyprinodon bifasciatus*, *Gambusia marshi*, and *Mexithauma quadripaludium* to meeting estimated P demand of primary producers in Río Mesquites

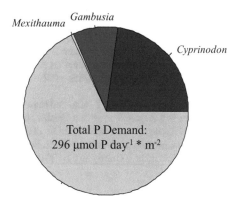

variable environments with minimal spring flow is limited (Carson et al. 2008; Carson 2009). Considering that *C. bifasciatus* contribute substantially to P recycling in springs of Cuatro Ciénegas, their loss could exacerbate P limitation. While *G. marshi* persist in environmentally variable environments, their P excretion rates are lower in these environments relative to those in springs. As a result, reduced spring flow resulting from groundwater extraction has the potential to substantially reduce P recycling by fishes. Such changes are likely already occurring in springs like Churince, where *C. bifasciatus* have been extirpated and *G. marshi* P excretion rates are reduced relative to those in other springs. The effects of these changes on the unique microbial communities of Cuatro Ciénegas is an important question, but one that unfortunately has not been examined in much detail. While much work, including the other chapters of this book, has investigated the function of these unique ecosystems, only a small fraction of this has considered the role animal consumers play (Garcia-Pichel et al. 2004; Elser et al. 2005a, b). More work on the effects of these and the other threatened animals of Cuatro Ciénegas will shed light on how these systems will change if groundwater levels continue declining around the basin.

References

Allgeier JE, Wenger SJ, Rosemond AD et al (2015) Metabolic theory and taxonomic identity predict nutrient recycling in a diverse food web. PNAS 112:E2640–E2647

Araújo MS, Langerhans RB, Giery ST et al (2014) Ecosystem fragmentation drives increased diet variation in an endemic livebearing fish of the Bahamas. Ecol Evol 4:3298–3308

Arnold ET (1972) Behavioral ecology of two pupfishes (Cyprinodontidae, genus Cyprinodon) from northern Mexico. Ph.D. dissertation, Arizona State University, Tempe

Atkinson CL, Vaughn CC, Forshay KJ et al (2013) Aggregated filter-feeding consumers alter nutrient limitation: consequences for ecosystem and community dynamics. Ecology 94:1359–1369

Atkinson CL, Capps KA, Rugenski AT et al (2017) Consumer-driven nutrient dynamics in freshwater ecosystems: from individuals to ecosystems. Biol Rev 92:2003–2023

Carson EW (2009) Threatened fishes of the world: *Cyprinodon bifasciatus* Miller 1968 (Cyprinodontidae). Environ Biol Fish 86:445–446

Carson EW, Dowling TE (2006) Influence of hydrogeographic history and hybridization on the distribution of genetic variation in the pupfishes *Cyprinodon atrorus* and *C. bifasciatus*. Mol Ecol 15:667–679

Carson EW, Elser JJ, Dowling TE (2008) Importance of exogenous selection in a fish hybrid zone: insights from reciprocal transplant experiments. Copeia 2008:794–800

Corman JR, Poret-Peterson AT, Uchitel A et al (2016) Interaction between lithification and resource availability in the microbialites of Río Mesquites, Cuatro Ciénegas, Mexico. Geobiology 14:176–189

Devine JA, Vanni MJ (2002) Spatial and seasonal variation in nutrient excretion by benthic invertebrates in a eutrophic reservoir. Freshw Biol 47:1107–1121

Dinger EC, Cohen AE, Hendrickson DA et al (2005) Aquatic invertebrates of Cuatro Ciénegas, Coahuila, Mexico: natives and exotics. Southwest Nat 50:237–281

Doughty CE, Roman J, Faurby S et al (2016) Global nutrient transport in a world of giants. PNAS 113:868–873

Elser JJ, Urabe J (1999) The stoichiometry of consumer-driven nutrient recycling: theory, observations, and consequences. Ecology 80:735–751

Elser JJ, Elser MM, MacKay NA et al (1988) Zooplankton-mediated transitions between N-and P-limited algal growth. Limnol Oceanogr 33:1–14

Elser JJ, Schampel JH, Garcia-Pichel F et al (2005a) Effects of phosphorus enrichment and grazing snails on modern stromatolitic microbial communities. Freshw Biol 50:1808–1825

Elser JJ, Schampel JH, Kyle M et al (2005b) Response of grazing snails to phosphorus enrichment of modern stromatolitic microbial communities. Freshw Biol 50:1826–1835

Garcia-Pichel F, Al-Horani FA, Farmer JD et al (2004) Balance between microbial calcification and metazoan bioerosion in modern stromatolitic oncolites. Geobiology 2:49–57

Garrett P (1970) Phanerozoic stromatolites: noncompetitive ecologic restriction by grazing and burrowing animals. Science 169:171–173

Hernández A, Espinosa-Pérez HS, Souza V (2017) Trophic analysis of the fish community in the Ciénega Churince, Cuatro Ciénegas, Coahuila. PeerJ 5:e3637

Hershler R (1984) The hydrobiid snails (Gastropoda: Rissoacea) of the Cuatro Cienegas basin: systematic relationships and ecology of a unique fauna. J Arizona Nevada Acad Sci 19:61–76

Kodric-Brown A (1988) Effect of population density, size of habitat, and oviposition substrate on the breeding system of pupfish (*Cyprinodon pecosensis*). Ethology 77:28–43

Lee ZMP, Poret-Peterson AT, Siefert JL et al (2017) Nutrient stoichiometry shapes microbial community structure in an evaporitic shallow pond. Front Microbiol 8:949

McIntyre PB, Flecker AS, Vanni MJ et al (2008) Fish distributions and nutrient cycling in streams: can fish create biogeochemical hotspots. Ecology 89:2335–2346

Meffe GK (1985) Life history patterns of Gambusia marshi (Poeciliidae) from Cuatro Cienegas, Mexico. Copeia 1985:898–905

Minckley WL (1962) Two new species of fishes of the genus Gambusia (Poeciliidae) from northeastern Mexico. Copeia 1962:391–396

Minckley WL (1969) Environments of the bolsón of Cuatro Cienegas, Coahuila, Mexico. Texas Western Press, University of Texas Press, El Paso

Minckley WL (1992) Three decades near Cuatro Ciénegas, México: photographic documentation and a plea for area conservation. J Arizona Nevada Acad Sci 26:89–118

Minckley WL, Cole GA (1968) Preliminary limnologic information on waters of the Cuatro Cienegas Basin, Coahuila, Mexico. Southwest Nat 13:421–431

Moody EK, Carson EW, Corman JR et al (2018) Consumption explains intraspecific variation in nutrient recycling stoichiometry in a desert fish. Ecology 99:1552–1561

Moody EK, Lozano-Vilano ML (2018) Predation drives morphological convergence in the Gambusia panuco species group among lotic and lentic habitats. J Evol Biol 31:491–501

Moody EK, Corman JR, Elser JJ et al (2015) Diet composition affects the rate and N: P ratio of fish excretion. Freshw Biol 60:456–465

Riding R (2000) Microbial carbonates: the geological record of calcified bacterial–algal mats and biofilms. Sedimentology 47:179–214

Small GE, Pringle CM, Pyron M et al (2011) Role of the fish Astyanax aeneus (Characidae) as a keystone nutrient recycler in low-nutrient Neotropical streams. Ecology 92:386–397

Souza V, Espinosa-Asuar L, Escalante AE et al (2006) An endangered oasis of aquatic microbial biodiversity in the Chihuahuan desert. PNAS 103:6565–6570

Sterner RW (1990) The ratio of nitrogen to phosphorus resupplied by herbivores: zooplankton and the algal competitive arena. Am Nat 136:209–229

Subalusky AL, Dutton CL, Rosi EJ et al (2017) Annual mass drownings of the Serengeti wildebeest migration influence nutrient cycling and storage in the Mara River. PNAS 114:7647–7652

Tobler M, Carson EW (2010) Environmental variation, hybridization, and phenotypic diversification in Cuatro Ciénegas pupfishes. J Evol Biol 23:1475–1489

Vanni MJ (2002) Nutrient cycling by animals in freshwater ecosystems. Annu Rev Ecol Syst 33:341–370

Vanni MJ, McIntyre PB (2016) Predicting nutrient excretion of aquatic animals with metabolic ecology and ecological stoichiometry: a global synthesis. Ecology 97:3460–3471

Vanni MJ, Boros G, McIntyre PB (2013) When are fish sources vs. sinks of nutrients in lake ecosystems? Ecology 94:2195–2206

Whiles MR, Hall RO, Dodds WK et al (2013) Disease-driven amphibian declines alter ecosystem processes in a tropical stream. Ecosystems 16:146–157

Chapter 12
How Do Agricultural Practices Modify Soil Nutrient Dynamics in CCB?

Yunuen Tapia-Torres, Pamela Chávez Ortiz, Natali Hernández-Becerra, Alberto Morón Cruz, Ofelia Beltrán, and Felipe García-Oliva

Contents

Abstract Increasing crop production to satisfy food demand of the growing population is one of the greatest challenges that we currently face. This has increased the rate at which natural ecosystems are transformed into agricultural systems. These land use changes, which are accompanied by agricultural practices such as the use of agrochemicals (pesticides and fertilizers), tillage, and irrigation, can lead to ecosystem degradation. Nationally, the state of Coahuila ranks seventh in production of alfalfa (1, 742,149 Mg), which is primarily used to feed the livestock in the country's largest milk-producing area, Comarca Lagunera in Torreón. The alfalfa produced at Cuatro Cienegas Basin (CCB) is transported to this region, therefore, effectively exporting the wetland water and threatening CCBs' sustainability.

For more than 10 years, our research group has been studying soil nutrient dynamics and soil bacterial biodiversity in CCB. Analysis of key soil attributes has allowed us to evaluate the drastic changes caused by land use change (native grassland into agricultural land) and to identify the soil processes that are vulnerable to management and, therefore, the factors that modify nutrient availability for the biota

Y. Tapia-Torres (✉) · N. Hernández-Becerra · A. M. Cruz
Escuela Nacional de Estudios Superiores Unidad Morelia, Universidad Nacional Autónoma de México, Morelia, México
e-mail: ytapia@enesmorelia.unam.mx

P. C. Ortiz · O. Beltrán · F. García-Oliva
Instituto de Investigaciones en Ecosistemas y Sustentabilidad, Universidad Nacional Autónoma de México, Morelia, México

© Springer International Publishing AG, part of Springer Nature 2018
F. García-Oliva et al. (eds.), *Ecosystem Ecology and Geochemistry of Cuatro Cienegas*, Cuatro Ciénegas Basin: An Endangered Hyperdiverse Oasis,
https://doi.org/10.1007/978-3-319-95855-2_12

153

and how these factors could increase soil degradation. Our findings can serve as the basis for implementation of land remediation practices in CCB, and the set of soil attributes and variables used in our studies may also be extrapolated to different ecosystems for the same purpose.

Keywords Alfalfa · Herbicides · Nitrification · Soil bacteria community · Soil nutrient dynamics

Introduction

Drylands constitute the most extensive terrestrial biome on the planet, covering more than one-third of the Earth's continental surface (Pointing and Belnap 2012). In Mexico, drylands constitute 40% of the total surface (Challenger 1998). Increasing global food demand expanded conversion of dryland ecosystems into agricultural production systems with a number of environmental impacts. Food production using management techniques such as the intensive use of fertilizers and irrigation by flooding the fields threatens dryland sustainability and also has a strong impact on soil processes, microorganisms, and vegetation.

Different types of agricultural practices affect the soil biota and could reduce genetic biodiversity (FAO 2013). In soils, bacteria and fungi are the most abundant and diverse group of microorganisms and play fundamental roles in many ecosystem processes such as biogeochemical cycles. Since the soil microbial community can respond rapidly to any disturbance, it may be strongly influenced by the land use change through changes in soil attributes. Therefore, soil biodiversity is threatened by unsustainable land use systems and management practices (FAO 2013).

Nowadays, agriculture has been characterized by its dependence on the use of agrochemicals, including pesticides. The soil is highly impacted by these compounds because it is more frequently exposed (Gianfreda and Rao 2011). When these compounds are applied constantly in the soil during agricultural activities, the presence of various bioactive molecules can cause negative effects in the microbial soil community and can change ecosystem processes, such as energy flow and nutrient dynamics (Wardle and Parkinson 1990).

Additionally, typical agricultural management systems for crops include inorganic fertilization. For example, phosphorus is added to crops at 50–300 kg P ha^{-1}, with applications often occurring regardless of the type of soil, the amount of P already in the soil, and the abundance and diversity of microorganisms that can increase soil P availability. In this chapter, we summarize the work developed for over 10 years in CCB with the objectives of understanding how agricultural management modifies the soil processes and, with this knowledge, of generating strategies for sustainable management of the land in this important Mexican desert ecosystem.

Soil Management for Alfalfa Production

Coahuila state ranks seventh in alfalfa production in Mexico with 57% of the area planted statewide (SDR 2012). The main destination of alfalfa produced in Cuatro Cienegas municipality is La Comarca Lagunera, one of the most important milk production sites in Mexico. Between 2001 and 2010, alfalfa encompassed more than 50% of the land under any crop production most of the time, and, during its peak for 2 years (2002 and 2004), this crop covers 100% of cultivated area (Beltrán 2017). Throughout the period analyzed, cultivation of alfalfa contributed the highest income from agricultural activity of the municipality (Beltrán 2017). Therefore, it is clear that alfalfa crop had the greatest economic influence as well as the largest planted area in the region in that period of time. However, this is changing as water access is more restricted and alfalfa less productive.

Historically, the region sustained vineyards as well as pomegranate and pecan trees that accessed water directly from the aquifer. As the deep aquifer diminished and the trees died (Souza et al. 2006), the diverse of crops related to human consumption diminished. Meanwhile, the dairy industry became more important in Torreon, taking the place of cotton, which was the most important crop in the region during most of the twentieth century. During the twenty-first century, agricultural lands in the CCB have been mainly used for the production of alfalfa (*Medicago sativa* L.), under flood irrigation and chemical fertilization (Fig. 12.1). The large volumes of water required by this crop have led to the desiccation of the main lakes and wetlands (DOF 2008). The amount of water used for the cropland irrigation in the CCB is 14 million m^3 yr^{-1} (DOF 2008). In the majority of cultivated plots, irri-

Fig. 12.1 Alfalfa (*Medicago sativa*) production under flood irrigation in the CCB

gation, inorganic fertilization, and alfalfa harvest occur once a month, but the alfalfa plants are replaced only every 3 years (Beltrán 2017). Unfortunately, after some years of cultivation, farm plots are abandoned due to fertility loss or salinization processes, which in most cases are not addressed because of the lack of economic means of the farmers (Beltrán 2017). For instance, there was a fourfold increase of the soil electrical conductivity in an abandoned plot compared to the cultivated one (3.4 and 15.6 mS m^{-1} for cultivated and abandoned plot, respectively), which suggests severe problems of soil salinization (Hernández-Becerra et al. 2016).

Changes in Soil Nutrient Dynamics After Agricultural Management

We assessed soil conditions across an agricultural gradient composed of three sites: a native grassland, an alfalfa plot, and a former agricultural field that had been abandoned for over 30 years (Hernández-Becerra et al. 2016). Our results indicate that the transformation of native grassland into alfalfa plots reduces bacterial diversity and induces drastic changes in soil nutrient dynamics. However, with the removal of farming, some of the soil characteristics analyzed in our work slowly recover toward their natural state (Hernández-Becerra et al. 2016).

The soil bacterial community of the native grassland was distributed among a higher number of phyla than the managed plots (12 and 9 respectively; Hernández-Becerra et al. 2016). However, the cultivated plots had a higher number of operational taxonomic units (OTUs; 92) than the native grassland (84 OTUs) and finally the abandoned plot (59 OTUs). Interestingly, we found that only a few OTUs were shared between sites. Four identical OTUs were identified in all three sites, and only the abandoned plot shared two OTUs with the other sites, but there were no OTUs shared between the agricultural plot and the native grassland (Fig. 12.2).

Additionally, the changes in the land use also modified the soil pH which decreased from 9 in native grassland to 7 in the alfalfa plot. As a consequence of contrasting

Fig. 12.2 Venn diagram representing the number of OTUs shared between sites and unique to each site (Hernández-Becerra 2014)

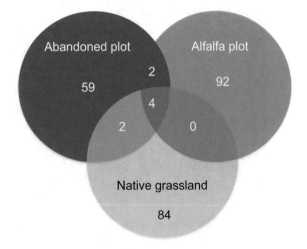

physical-chemical conditions, bacterial diversity changed, and the soil biogeochemical processes were modified by the loss of some key functional microbial groups. On one hand, the nitrification processes strongly increased in the soils of the alfalfa crops as well as in abandoned plots, likely as a consequence of increase N availability due to fertilization. Moreover, cyanobacteria were lost from the microbial crust that predominated in the native soils. With the loss of cyanobacteria as well as *Rhizobium* (alfalfa is a legume that is able to nodulate and fix nitrogen), the soil has a reduced capability to fix the atmospheric nitrogen, thus increasing the need for fertilizers. As a consequence, nitrogen is more vulnerable to losses on the soil of sites under management.

On the other hand, phosphorus, a fundamental element for all life including crops, is very vulnerable to management. Inorganic phosphate is the only form of phosphorus that cells can use directly for incorporation into essential biological molecules, such as DNA and RNA, and into the phospholipid bilayer in membranes (Tapia-Torres and Olmedo-Álvarez 2018). Soil processes associated with phosphorus availability are very vulnerable to management due to the low phosphorus soil content naturally present in the CCB. We observed decreased abundances of bacteria associated with phosphorus transformation likely attributable to impacts of farming (Tapia-Torres et al. 2016). For example, phosphatase activity decreased significantly in the cultivated plots in relation to the native grassland soil (Hernández-Becerra et al. 2016). However, we also identified a modification in organic nutrient availability as a consequence of land use change. Abandoned land was limited by water and dissolved organic nitrogen. The low amount of dissolved organic matter promoted nitrification, which is mediated by autotrophic microorganisms (Fig. 12.3). As a conclusion, soil nutrient dynamics appeared and

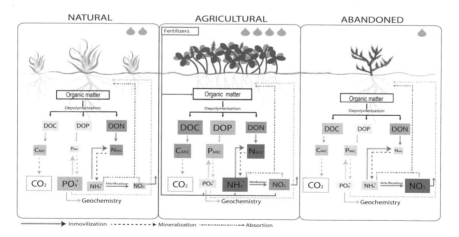

Fig. 12.3 Conceptual flow model of soil C, N, and P dynamics across an agricultural gradient. For each fraction, the dimension of the box represents the size of the pool among sites. Significant differences ($p < 0.05$) of fractions between sites are shown with the contrast of color shading. The agricultural site had greater availability of water and dissolved organic carbon (DOC); these conditions stimulated mineralization processes. In contrast, both water and DOC limited sites without current management. At the abandoned site, these conditions promoted nitrification; meanwhile, N immobilization within the microbial biomass was favored in the natural site (Hernández-Becerra 2014)

the structure of soil bacterial community to be strongly affected by present-day farming practices at CCB.

Herbicides and Its Effects on the Soil Nutrient Dynamics

In the CCB a large variety of pesticides are used to maintain the crops free of unwanted organisms. Among the pesticides, the use of herbicides stands out. It has been reported that the herbicides Select™ (cyclohexadiene), Cedrus™ (cyclohexadiene), and Pivot™ (imazethapyr; Hernández-Becerra 2014) are used in the alfalfa plots of CCB. Where the alfalfa crops have been replaced by nopal (prickly pear) cultivation, glyphosate herbicides (n-phosphonomethyl glycine) are used (Chavez-Ortiz 2017).

Glyphosate is a chemical compound characterized by its phosphonate bond (C-P) and is one of the most used herbicides in recent years worldwide (Duke and Powles 2008). This molecule, due to its three functional groups (phosphonic acid, carboxylic acid, and secondary amine; Damonte et al. 2007) has the ability to adsorb onto soil clays and hydroxides (Sprankle et al. 1975), competing with the ortho-phosphates (HPO_4^{3-}) for sorption sites in the soil. Studies carried out on CCB agricultural soils have reported that soil minerals have a high capacity to occlude P. The more P is added to the soil, the more P is occluded (Perroni et al. 2014; Chávez-Ortiz 2017). In addition, the presence of higher concentrations of organic compounds favors P occlusion (Chavez-Ortiz 2017). Indeed, a commercial formulation of glyphosate had the capacity to significantly decrease P occlusion in the soil and the rate at which this element is occluded in the soil (Chavez-Ortiz 2017). This reduction can be attributed to the other compounds that the herbicide contains as adjuvants that reduce chemical reactions in the soil. The lower P occlusion in soil treated with the commercial formulation could mean a greater occlusion of the herbicide in the soil and a longer residence time of this compound in the soil.

Because of the effects that glyphosate can produce on the environment and other living beings, it is important to find a way to reduce concentrations of this compound in the soil. Strategies to mitigate the problems of contamination by pesticides include bioremediation techniques that use microorganisms to degrade polluting compounds (Iranzo et al. 2001). Due to the conditions of P limitation of the CCB, microorganisms at CCB likely have evolved a variety of genetic tools to metabolize P-containing compounds. To assess this, several studies have been carried out to find microorganisms native to CCB that can degrade glyphosate, using it as source of P. For example, Morón-Cruz (2014) used 45 bacterial isolates from the soils of Churince and Pozas Azules sites at CCB that had previously been identified in the work of Tapia-Torres et al. (2016) to test for their ability to use the compound n-phosphonomethyl glycine (glyphosate) as the sole source of P. As a result of this experiment, ten isolates with this capacity belonging to the genera *Bacillus*, *Pantoea*, and *Pseudomonas* were obtained. In addition, Chávez-Ortiz (2017) conducted laboratory experiments using soil from nopal plots and from an abandoned farm and evaluated potential soil C mineralization by adding 96% pure glyphosate and a commercial formula of the herbicide to understand how the addition of glyphosate affects soil microbial activity. Lower potential mineralization rate was observed in soils receiving the purified glyphosate, perhaps due to the fact that glyphosate can affect soil microorganisms by inhibiting the shikimic acid pathway, which is carried out not only by plants but also by some bacteria and fungi (Bentley and Haslam 1990; Ratcliff et al. 2006). Inhibition of this metabolic route affects protein synthesis by preventing the formation of aromatic amino acids (Bentley and Haslam 1990; Franz et al. 1997). However, later in this same study (Chavez-Ortiz 2017), bacteria from soil with different treatments (glyphosate and commercial herbicide) were isolated, and the abilities of each of the isolates to grow using 96% pure glyphosate or the commercial formulation of the herbicide as the only source of P were tested. In this study, 88% of the isolates managed to grow using reagent-grade glyphosate, while only 30% could use the commercial herbicide as a sole P source. The growth of bacteria using glyphosate as a source of P, an anthropogenic phosphonate, indicates that most of the bacteria isolated from both the agricultural and the abandoned sites have the enzymes necessary for mineralization of phosphonates. These might involve the enzymatic complex C-P lyase or the enzyme phosphonatase (Pipke and Amrhein 1988; Sviridov et al. 2012, 2015). However, not all bacteria had the ability to survive in culture media with the commercial formulation of the herbicide, possibly due to toxic characteristics of other ingredients found in its formulation, such as surfactants (e.g., polyoxyethylene tallow amine).

Conclusion

Soil microorganisms are key for the understanding of soil processes. The set of microbial genes with different specific functions regulate soil biogeochemical processes. However, soil microorganisms (bacteria and fungi) have short generation times and therefore respond rapidly to changes in the physical-chemical conditions of the environment in which they live. Land use change drastically modifies these physical-chemical conditions due to the use of agrochemicals and irrigation, altering the diversity and abundance of the native soil microbiota and therefore altering soil genetic resources. As a consequence of these environmental perturbations, soil is very vulnerable to degradation. However, soil diversity itself helps lend resilience to land use change, as native genetic resources of the microbial community can be the answer to soil reclamation.

Our findings suggest that the most sensitive processes to soil management are related to the P recycling due to the low P content naturally present in the CCB. Reflecting this P limitation, soil and sediment bacteria in CCB are able to break down and use P forms in different oxidation states and, therefore, contribute to ecosystem P cycling. The various strategies for P utilization are distributed between and within different taxonomic lineages, suggesting a dynamic movement of P utilization traits among bacteria in microbial communities. Likewise, it was possible to determine that, with the removal of agricultural practices, some of the soil characteristics slowly recover their natural state. These findings can serve as the basis for the implementation of land remediation practices in the Cuatro Cienegas Valley. We also suggest that the set of soil attributes used in our studies can also be applied to different ecosystems for the same purpose.

References

Beltrán O (2017) Dinámica de nutrientes del suelo bajo cultivo intensivo de alfalfa en la región ganadera del valle de Cuatro Ciénegas, Coahuila. Master thesis, Ciencias Biológicas, UMSNH

Bentley R, Haslam E (1990) The shikimate pathway—a metabolic tree with many branches. Crit Rev Biochem Mol Biol 25:307–384

Challenger A (1998) Utilización y conservación de los ecosistemas terrestres de México, Pasado, presente y Futuro. Comisión Nacional para el Conocimiento de los Ecosistemas y Uso De La Biodiversidad, México, D.F

Chávez-Ortiz P (2017) Efecto del uso del glifosato en la dinámica de nutrientes y actividad microbiana de suelos agrícolas en el valle de Cuatro Ciénegas, Coahuila. Master thesis, Ciencias Biológicas, UNAM

Damonte M, Sánchez RMT, dos Santos Afonso M (2007) Some aspects of the glyphosate adsorption on montmorillonite and its calcined form. Appl Clay Sci 36:86–94

DOF. Diario Oficial de la Federación (2008) Acuerdo por el que se dan a conocer los estudios técnicos del Acuífero 0528 Cuatrociénegas y se modifica los límites y plano de localización que respecto del mismo se dieron a conocer en el Acuerdo por el que se dan a conocer los límites de 188 acuíferos de los estados Unidos mexicano, los resultados de los estudios realizados para determinar su disponibilidad media anual de agua y sus planos de localización. Poder Ejecutivo. Secretaria de Medio Ambiente y Recursos Naturales (SEMARNAT). Primera Sección. Junio

Duke SO, Powles SB (2008) Mini-review Glyphosate: a once-in-a-century herbicide. Pest Manag Sci 64:319–325

FAO (2013) FAO statistical year book. World food and agriculture. Food and Agriculture Organization of the United Nations, Rome

Franz JE, Mao MK, Sikorski JA (1997) Glyphosate: a unique global herbicide (American Chemical Society Monograph 189). American Chemical Society, Washington, DC, 653p

Gianfreda L, Rao MA (2011) The influence of pesticides on soil enzymes. In: Shukla G, Varma A (eds) Soil enzymology. Springer, Berlin/Heidelberg, pp 293–312

Hernández-Becerra N (2014) Dinámica de C, N y P y composición de la comunidad bacteriana del suelo de un gradiente de manejo agrícola en el Valle de Cuatro Ciénegas, Coahuila. Bachelor thesis, UNAM

Hernández-Becerra N, Tapia-Torres Y, Beltrán-Paz O et al (2016) Agricultural land-use change in a Mexican oligotrophic desert depletes ecosystem stability. PeerJ 4:e2365

Iranzo M, Sain-Pardo I, Boluda R et al (2001) The use of microorganisms in environmental remediation. Ann Microbiol 51:135–143

Morón-Cruz, JA (2014) Degradación de N-fosfonometil glicina (glifosato) por bacterias edáficas de Cuatro Ciénegas, Coahuila. Bachelor thesis, Instituto Tecnológico de Morelia

Perroni Y, García-Oliva F, Tapia-Torres Y et al (2014) Relationship between soil P fractions and microbial biomass in an oligotrophic grassland-desert scrub system. Ecol Res 29:463–472

Pipke R, Amrhein N (1988) Degradation of the phosphonate herbicide glyphosate by Arthrobacter atrocyaneus ATCC 13752. Appl Environ Microbiol 54:1293–1296

Pointing SB, Belnap J (2012) Microbial colonization and controls in dryland systems. Nat Rev Microbiol 10:551e563

Ratcliff AW, Busse MD, Shestak CJ (2006) Changes in microbial community structure following herbicide (glyphosate) additions to forest soils. Appl Soil Ecol 34:114–124

Secretaría de Desarrollo Rural (2012) Programa Estatal de Desarrollo Rural 2011–2017. Coahuila de Zaragoza Gobierno del Coahuila, Secretaría de Desarrollo Rural Saltillo, Coahuila

Souza V, Espinosa-Asuar L, Escalante AE, Eguiarte LE, Farmer J, Forney L, Lloret L, Rodriguez-Martinez JM, Soberon X, Dirzo R, Elser JJ (2006) An endangered oasis of aquatic microbial biodiversity in the Chihuahuan desert. Proc Nat Acad Sci 103(17):6565–6570

Sprankle P, Meggitt WF, Penner D (1975) Adsorption, mobility, and microbial degradation of glyphosate in the soil. Weed Sci 23:229–234

Sviridov AV, Shushkova TV, Zelenkova NF et al (2012) Distribution of glyphosate and methylphosphonate catabolism systems in soil bacteria Ochrobactrum anthropi and Achromobacter sp. Appl Microbiol Biotechnol 93:787–796

Sviridov AV, Shushkova TV, Ermakova IT et al (2015) Microbial degradation of glyphosate herbicides (Review). Appl Biochem Microbiol 51:188–195

Tapia-Torres Y, Rodríguez-Torres MD, Elser JJ et al (2016) How to live with phosphorus scarcity in soil and sediment: lessons from bacteria. Appl Environ Microbiol 82:4652–4662

Tapia-Torres Y, Olmedo-Álvarez G (2018) Life on phosphite: a Metagenomics tale. Trends Microbiol 26:170–172

Wardle DA, Parkinson D (1990) Influence of the herbicide glyphosate on soil microbial community structure. Plant Soil 122:29–37

Index

© Springer International Publishing AG, part of Springer Nature 2018   163
F. García-Oliva et al. (eds.), *Ecosystem Ecology and Geochemistry of Cuatro Cienegas*, Cuatro Ciénegas Basin: An Endangered Hyperdiverse Oasis,
https://doi.org/10.1007/978-3-319-95855-2